QC
225.15
F67

Ford

Introduction to acoustics

150896

DATE DUE	DATE RETURNED
DEC 17 1992	DEC. 1 6 REC'D

D0820318

FACULTY		
OCT 2 6 1980		
NOV 1 7 1980	Renewal	
Oct 11-26-80		

NEW MEXICO
EASTERN UNIVERSITY
E N M U
LIBRARY

Introduction to Acoustics

Introduction to Acoustics

R. D. Ford

Senior Lecturer, Department of Pure and Applied Physics
The University of Salford, Salford, Lancashire

ELSEVIER PUBLISHING COMPANY LIMITED

AMSTERDAM — LONDON — NEW YORK

1970

ELSEVIER PUBLISHING COMPANY LTD
BARKING, ESSEX, ENGLAND

ELSEVIER PUBLISHING COMPANY
335 JAN VAN GALENSTRAAT, P.O. BOX 211 AMSTERDAM
THE NETHERLANDS

AMERICAN ELSEVIER PUBLISHING COMPANY INC.
52 VANDERBILT AVENUE, NEW YORK, N.Y. 10017

0444-20078-9

LIBRARY OF CONGRESS CATALOG CARD NUMBER 75-125568

WITH 67 ILLUSTRATIONS AND 11 TABLES

© COPYRIGHT 1970 ELSEVIER PUBLISHING COMPANY LTD

All rights reserved. No part of this publication may be reproduced, stored in a
retrieval system, or transmitted in any form or by any means, electronic,
mechanical, photocopying, recording, or otherwise without the prior written
permission of the publisher, Elsevier Publishing Company Ltd, Ripple Road,
Barking, Essex, England

Printed in Great Britain by Galliard Limited, Great Yarmouth, England

QC
225.15
F67

Contents

150896

Preface

The study of acoustics is a comparatively new discipline in universities and the supply of textbooks suitable for a short course is very limited. This book is largely based on the author's experience of lecturing to final year undergraduate students and is quite sufficient to give a good grounding in the subject. No previous knowledge of acoustics is assumed but some mathematical ability is required because complex algebra is used extensively. No apology is offered for this because complex algebra is a common mathematical tool with which modern scientists and technologists must be familiar and which would certainly be needed if the study of acoustics were to be pursued beyond the limits of this book.

The contents are intentionally restricted to the audio-frequency range because ultrasonics, although sometimes classified as acoustics, is really a subject in its own right with quite different technological applications. The aim of this book is to explain the fundamental principles of acoustics. Applications are not stressed but it is thought that study of a particular problem, using books which are more applied in nature, will follow quite easily. Normal acoustic terminology is used throughout and the first chapter should be studied until the reader is thoroughly familiar with decibels and frequency bands.

Similarities between parts of this book and 'Fundamentals of Acoustics' by L. E. Kinsler and A. R. Frey will be noticed by anyone already well versed in the subject. This is acknowledged but any changes in mathematical approach would be undesirable. Students wishing to pursue the subject of acoustics beyond the limits of this book will find the more advanced and comprehensive book mentioned above invaluable.

Thanks are due to Professor P. Lord for reading the manuscript and for offering numerous helpful criticisms and to Mrs. J. Y. Ford for typing the manuscript and for proof-reading.

ix

CHAPTER 1

Introduction

1.1 Sound Waves

We are familiar with sound waves travelling through air but give little thought to the physical process which is occurring. In fact sound is capable of travelling through any medium and it is simply the molecular transfer of motional energy. The obvious conclusion is that sound cannot pass through a vacuum and this is effectively proved in junior schools by placing a ringing bell in an evacuated chamber.

The viscosity of gases is so low that shear waves cannot be supported and disturbances can only travel as compressional waves. This is also true for many liquids, certainly for water which, after air, is technologically the most important medium in acoustics. Compressional waves can travel through solids and this is the most significant wave in long thin rods, but in the general case both shear and bending waves can also be supported and so the situation is much more complex. Waves in solids and viscous liquids are, however, of greater significance in the field of ultrasonics and only compressional waves will be considered in this book.

A compressional wave can perhaps be visualised most easily in terms of a small pressure disturbance (or stress in a solid) which communicates itself to the surrounding medium. Creating ripples in a pond by throwing a stone could be cited as a visual example although it must be realised that these are surface waves and not compressional waves. In three dimensions it is to be expected that the disturbance will gradually diminish as it travels outwards since the initial amount of energy is gradually spreading itself over a larger and larger area. If the disturbance is confined to one dimension, as

1

in a tube of fluid or in a thin solid rod, then it does not diminish as it travels through the medium, except for a negligibly small amount of energy which is dissipated by the movement of the molecules and along the sides of the tube or rod. Some motion of the molecules about their mean positions is, of course, associated with the pressure disturbance. This motion is dependent upon the magnitude of the disturbance and must not be confused with the rate at which the disturbance travels through the medium. The precise relationship between pressure and the motion is important because it is often a known movement which gives rise to the disturbance in the first place. The motion is along an axis parallel with the direction in which the disturbance is propagating and compressional waves are sometimes referred to as longitudinal waves for this reason. In solids the motion is described in terms of the strain (or displacement) but in fluids it is customary to describe the local molecular motion in terms of the particle velocity although displacement could be used.

1.2 Velocity of Sound

The disturbances travel through the medium at a rate determined by the speed at which the molecules can transfer energy from one to another. For small disturbances this is a constant depending only on the physical properties of the medium. For large disturbances, however, the speed is also influenced by the size of the disturbance and this is referred to as non-linear propagation. Fortunately the majority of sound waves are not so large and it is sufficient here to study linear propagation.

It will be shown in the next chapter that the velocity of sound in a gas may be expressed in the form

$$c = \left(\frac{\gamma P_0}{\rho_0}\right)^{1/2} \tag{1.1}$$

γ is the ratio of specific heats of the gas, P_0 is the mean pressure and ρ_0 is the mean density. γ appears in the equation because the fluctuations associated with the propagation of some disturbance are so rapid that there is no time for any heat transfer and so the relationship between fluctuating pressure and fluctuating density is adiabatic. The ratio P_0/ρ_0 is very nearly constant for most gases and the velocity of sound is independent of pressure. We may, however, write $P/\rho = RT/M$ where R is the gas constant, M is the molecular weight and T is the absolute temperature in $°K$ and then equation (1.1) becomes

$$c = (\gamma RT/M)^{1/2} \tag{1.2}$$

So the velocity of sound in a gas is proportional to the square root of the absolute temperature.

For liquids it is much more difficult to develop an equation giving the velocity of sound although in general it may be said that

$$c = \left(\frac{B_a}{\rho_0}\right)^{1/2} \tag{1.3}$$

where B_a is the adiabatic bulk modulus of elasticity and ρ_0 is the density. There are no simple relationships between these parameters and all three vary with temperature so no general conclusions can be drawn except that the velocity increases by a small amount with an increase of either pressure or temperature.

The velocity of longitudinal waves in solid rods is easy to derive and it is found that

$$c = \left(\frac{E}{\rho_0}\right)^{1/2} \tag{1.4}$$

where E is Young's modulus of elasticity and ρ_0 is the density of the material. The velocity is largely independent of both pressure and temperature.

A thin rod is not restricted laterally and as the wave progresses it is accompanied by a small change of cross-sectional area. This is known as the Poisson effect. If the solid is in bulk form then it is restricted from moving laterally, which increases the stiffness and the velocity of sound. The bulk modulus of elasticity must be substituted for Young's modulus in equation (1.4).

Table A.1, in the appendix at the end of the book, lists the velocity of sound in some common substances at 20°C.

1.3   Root Mean Square Pressure

In a fluid, or even in a solid, there is always a steady component of pressure. When the disturbance which we have been discussing passes a certain point it will appear as a small pressure fluctuation about the mean. This is illustrated in Fig. 1.1 for a sound wave in air at atmospheric pressure.

The total pressure at any time $= P_0 + p(t)$ where $p(t)$ represents the sound wave. We have no interest in the mean pressure component, P_0, but we need to have some way of defining the strength of the fluctuating component $p(t)$. The mean energy associated with the sound wave is its most fundamental feature and since the energy is

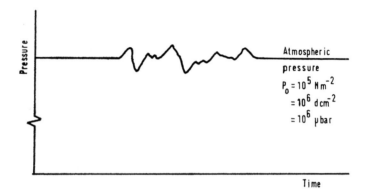

Fig. 1.1.   *Fluctuating pressure associated with the passage of a sound wave.*

proportional to the square of the pressure it is conventional to use the mean square pressure,

i.e.
$$\bar{p}^2 = \frac{1}{T} \int_0^T p(t)^2 \, dt \tag{1.5}$$

where T represents the time period of interest. Then the strength of the pressure fluctuation is the square root of the mean square pressure, *i.e.* $(\bar{p}^2)^{1/2}$, which is often written simply as $\bar{p}$ and is called the root mean square (or rms) pressure.

If the sound wave is harmonic the pressure fluctuation at a point can be represented by $p = a \cos \omega t$.

Then
$$\bar{p}^2 = \frac{1}{T} \int_0^T a^2 \cos^2 \omega t \, dt$$

$$= \frac{a^2}{2}$$

and
$$\bar{p} = \frac{a}{2^{1/2}} = 0 \cdot 707a \tag{1.6}$$

In this case there is a simple relationship between the rms pressure and the peak pressure which is worth remembering since many problems may be understood by assuming a harmonic form for the solution.

Human hearing is discussed in more detail in Chapter 7 but it should be noted here that the range of rms pressure fluctuations in which we are interested for audio-acoustics extends from about 2×10^{-5} Nm^{-2} to 20 Nm^{-2}. The former corresponds approximately to the minimum pressure fluctuation discernible to the human ear and the latter to the point at which sound becomes a painful sensation. Remembering that atmospheric pressure is $10^5 Nm^{-2}$ it is evident that even the loudest sounds represent only very small fluctuations of pressure about the steady value.

1.4 Decibels

In acoustics we are usually more concerned with multiplying a known pressure by a given factor than adding or subtracting one pressure from another. One example of this would be altering the volume control on a hi-fi set to increase the voltage applied to the loudspeaker by a factor of perhaps 5 or 10 resulting in the fluctuating sound pressure also being increased by a factor of 5 or 10. Another example might be a wall between two rooms which always permits, say, 1% of the sound pressure on one side to pass through to the other side irrespective of the magnitude of the original sound wave.

The same situation occurs in electrical engineering and there the multiplicative or divisive process has been changed into an additive or subtractive process by the simple expedient of taking logarithms. For example, if
$$W_1 = W_2 \times n$$
then
$$\log_{10} W_1 = \log_{10} W_2 + \log_{10} n$$

This idea was originally applied to power or energy and the unit applied to the logarithmic values was the Bel. Thus if W_1 and W_2 in the above equation represent two values of power and if $n = 10$ the difference between $\log_{10} W_1$ and $\log_{10} W_2$ is $\log_{10} 10$ which is 1 Bel. This particular scale proved a little inconvenient in that decimal points frequently had to be used but this is overcome by multiplying the equation throughout by 10,

i.e. $$10 \log_{10} W_1 = 10 \log_{10} W_2 + 10 \log_{10} n \qquad (1.7)$$

The unit changes from Bel to decibel (abbreviated to dB) and 1 Bel equals 10 decibels in the same way that 1 metre equals 10 decimetres. In the example above the difference between $10 \log_{10} W_1$ and $10 \log_{10} W_2$ is now 10 dB and if $n = 100$ then the difference is 20 dB.

If we wish to apply the concept to electrical voltage we can do so by writing $W = V^2/R$ where R is a constant resistance. Then equation (1.7) becomes

$$10 \log_{10} V_1^2/R = 10 \log_{10} V_2^2/R + 10 \log_{10} n$$

The constant R cancels out and the square term may be taken outside the logarithm, so giving

$$20 \log_{10} V_1 = 20 \log_{10} V_2 + 20 \log_{10} n^{1/2} \qquad (1.8)$$

Equation (1.8) has been derived directly from equation (1.7), so the unit is still the decibel and if $V_1 = V_2 \times 10^{1/2}$ (corresponding to $W_1 = W_2 \times 10$) the difference between $20 \log_{10} V_1$ and $20 \log_{10} V_2$ is still 10 dB.

Decibels then are simply another way of expressing a ratio and they can be used in acoustics just as easily as in electrical engineering. If we suppose that acoustical power is analogous to electrical power and acoustic pressure is analogous to electrical voltage, we may substitute pressure for voltage in equation (1.8), giving

$$20 \log_{10} p_1 = 20 \log_{10} p_2 + 20 \log_{10} n^{1/2} \qquad (1.9)$$

Table 1.1 lists the decibel values for a few simple ratios and in practice almost any ratio can be deduced from this table. For example, a pressure ratio of 6 can be written 2×3 which is equivalent to $6 + 10 = 16$ dB. If the ratio is a reciprocal of the numbers listed, *e.g.* if the ratio is $\frac{1}{2}$ rather than 2, then the difference is numerically the same but it takes a negative sign. So a pressure ratio of 5 could be written $10 \times \frac{1}{2}$ which is equivalent to $20 - 6 = 14$ dB. It will be noted that the decibel values in the final column of Table 1.1 have been rounded off to the nearest whole number. This is common since the accuracy associated with fractions of decibels is rarely warranted.

TABLE 1.1

DECIBEL VALUES CORRESPONDING TO A FEW SIMPLE RATIOS

Power ratio n	Voltage or pressure ratio $n^{1/2}$	Difference in dB $= 10 \log_{10} n$ or $20 \log_{10} n^{1/2}$
2	1·414	3
4	2	6
10	3·16	10
100	10	20

The common use of the decibel in acoustics and electrical engineering is of particular significance when both fields combine. Electroacoustics generally deals with transducers and the associated electrical equipment and controls on amplifiers and meters are frequently graduated in decibels. Power amplifiers produce an electrical voltage which is converted to a sound pressure by a loudspeaker and microphones convert a sound pressure to an electrical voltage which may then appear on a meter. The efficiency is low but there is a direct linear relationship between pressure and voltage in both cases. So if the voltage alters by a given number of decibels, the pressure alters by the same number of decibels.

1.5 Sound Pressure Level

The last section described how decibels could be used for expressing the ratio of two quantities. Equation (1.9) could be written

$$20 \log_{10} n^{1/2} = 20 \log_{10} \left(\frac{p_1}{p_2}\right)$$

If p_2 were given an absolute value and $20 \log_{10} n^{1/2}$ were known then p_1 would also have an absolute value. In audio-acoustics, p_2 is frequently given the absolute rms value of 2×10^{-5} Nm^{-2} which is the minimum sound pressure fluctuation discernible to the human ear. Consequently, p_1 must also become an rms pressure expressed in Nm^{-2}. $20 \log_{10} n^{1/2}$ is still in decibels but is now defined precisely as the sound pressure level (SPL).

i.e.
$$\text{SPL} = 20 \log_{10} \frac{\bar{p}_1}{2 \times 10^{-5}} \, \text{dB} \qquad (1.10)$$

In terms of sound pressure level the sensitivity of the ear ranges from 0 dB (2×10^{-5} Nm^{-2}) to 120 dB (20 Nm^{-2}). In fact the extremes are rarely encountered and everyday levels vary between about 35 and 90 dB.

The reference pressure of 2×10^{-5} Nm^{-2} is strictly a convention for audio-acoustics. Any value could be taken and in underwater acoustics a reference pressure of 10^{-1} Nm^{-2} is generally used. Therefore, to avoid any possible confusion, it would be correct to quote the sound pressure level as being so many decibels relative to the particular reference pressure which has been chosen.

1.6 Intensity Level

A sound wave travelling in one dimension along a tube contains energy and it will be shown in the next chapter that the relationship between the average intensity, I, that is the rate of transfer of energy per unit cross-sectional area, and the rms pressure is

$$I = \frac{\bar{p}^2}{\rho_0 c} \qquad (1.11)$$

where ρ_0 is the density of the medium and c is the velocity of sound in the medium. The product $\rho_0 c$ for some substances is listed in Table A.1 in the appendix at the end of the book. The same expression applies to sound waves radiating outwards in two or three dimensions provided that the point considered is far from the source.

Thinking in terms of acoustic intensity in air waves it is possible to rewrite equation (1.10) in the form

$$\text{SPL} = 10 \log_{10} \frac{\bar{p}_1^2/\rho_0 c}{(2 \times 10^{-5})^2/\rho_0 c} \text{ dB}$$

$\bar{p}_1^2/\rho_0 c$ is the intensity I, associated with the sound wave of pressure $\bar{p}_1$ and $(2 \times 10^{-5})^2/\rho_0 c$ is the intensity associated with the sound wave of reference pressure $2 \times 10^{-5} \, Nm^{-2}$. Substituting the value of $415 Nm^{-3}$ sec, the reference intensity is very nearly $10^{-12} \, Wm^{-2}$. So the equation may be written in terms of intensity and the intensity level (IL) is defined as follows.

$$\text{IL} = 10 \log_{10} \frac{I_1}{10^{-12}} \text{ dB} \qquad (1.12)$$

This is most convenient because, since the reference intensity corresponds almost exactly to the reference pressure, the intensity level in dB of a sound wave in air will be numerically equal to the sound pressure level in dB and the two can be interchanged at will.

In substances other than air the numerical relationship between pressure and intensity will be different and some care must be exercised. In water, for example, the reference intensity would need to be about $6.7 \times 10^{-9} \, Wm^{-2}$ and the reference pressure $10^{-1} \, Nm^{-2}$ for the intensity level and the sound pressure level to be numerically equal.

1.7 Contributions from Several Sources

If sound is being produced by more than one source then each will contribute to the sound pressure level or intensity level at any point. This is the one instance in acoustics when the use of decibels can be a little inconvenient.

First there is the possibility that the sources are coherent, that is they are all producing the same wave form as a function of time. This could easily be achieved by feeding two or more loudspeakers from the same amplifier. The situation is directly analogous to optical interference and it is necessary to sum the pressure fluctuations at the point of interest taking account of the phase relationships between the different waves. What usually happens is that at some points the waves all arrive in phase and add together so producing a considerable increase in pressure while at other points the waves arrive out of phase cancelling each other out and perhaps producing a zero pressure. In the simplest case of two coherent waves of equal pressure the sum can vary from zero at some points to a doubling of pressure at other points. Doubling the pressure corresponds, of course, to increasing the sound pressure level by 6 dB. Attempts have been made to apply the concept of pressure cancellation to noise control problems by locating a loudspeaker next to the offending noise source and producing the same amount of noise but exactly out of phase. At points equidistant from both sources there should theoretically be complete cancellation but usually the primary source is too large for this simple solution and at other points there would, anyway, be an increase in noise level. The idea has therefore met with no success in this particular application.

If the sources are not coherent then there is no possibility of a regular interference pattern being produced and the energy being distributed in a particular way. It is correct in these circumstances to add the intensities from the various sources at any point. The total intensity is given by

$$I_T = I_1 + I_2 + I_3 + \cdots$$

Usually, however, it is the intensity levels of the different sources which are known, *i.e.* L_1, L_2, etc., and then it is necessary to divide each by 10 and take the antilogarithm before adding to obtain the total intensity. The reference intensity is always the same and can be omitted. The total intensity level is finally given by

$$L_T = 10 \log_{10} [10^{L_1/10} + 10^{L_2/10} + 10^{L_3/10} + \cdots] \quad (1.13)$$

If the intensity level from each of N sources is the same then equation (1.13) simplifies to

$$L_T = 10 \log_{10} [N \times 10^{L_1/10}]$$
$$= L_1 + 10 \log_{10} N$$

The intensity level with two sources is 3 dB higher than with one source, 5 dB higher with three sources and 6 dB higher with four sources. It is necessary to double the number of sources for each increase of 3 dB.

Another special case that can be considered is when there are only two sources. Then equation (1.13) reduces to

$$L_T = 10 \log_{10} [10^{L_1/10} + 10^{L_2/10}]$$

$$= L_1 + 10 \log_{10} [1 + 10^{L_2/10}/10^{L_1/10}] \qquad (1.14)$$

Assuming that $L_1 \geqslant L_2$ then the last term in equation (1.14) is always $\leqslant 3$ dB. Figure 1.2 shows this last term ($= L_T - L_1$) plotted

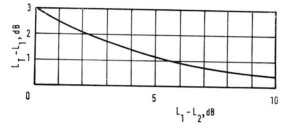

Fig. 1.2. Chart for adding together two intensity or sound pressure levels.

as a function of the difference of the two individual levels ($L_1 - L_2$). There is little point in carrying out the calculations in equation (1.14) more than once and the figure can be used for adding any two intensity levels. More than two can be added by a process of successive summation.

1.8   Frequency and Frequency Bands

The frequency of a sound is as important as its level, because the sensitivity of the ear varies with frequency; the sound insulation of a wall varies with frequency; the attenuation of a silencer varies with frequency; the ability of a sound wave to bend around corners varies with frequency and so on. First it must be stated that the frequency range for audio-acoustics extends from about 20 Hz to 15 kHz. Below 20 Hz the pressure fluctuations are felt by the whole body as much as by the ears although the level has to be very high in order to produce any sensation. These low frequencies are sometimes referred to as the infrasonic region. The ultrasonic region starts at about 15 kHz and the average human ear is totally insensitive to such high frequencies.

A pure tone of sound has a simple harmonic pressure fluctuation of constant frequency and amplitude. It would be represented by a single line on a graph of sound pressure level versus frequency, as in Figure 1.3 (a).

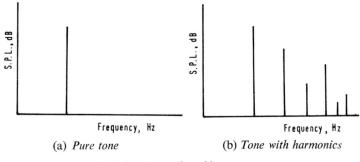

(a) *Pure tone*   (b) *Tone with harmonics*

Fig. 1.3.   Examples of line spectra.

A musical instrument does not produce a pure tone, but rather a combination of tones called partials. The lowest frequency partial is always referred to as the fundamental. The frequencies of the higher partials may be related integrally to the fundamental, as in a stringed instrument, in which case they are called harmonics or there may be no simple relationship, as with a bell, in which case they are called anharmonics. In either case the total wave form is not a pure sinusoid but it can be Fourier analysed into its several components each of which is a sinusoid. The strengths will vary and a graphical representation would consist of a series of vertical lines as in Figure 1.3 (b).

Noise, as produced by a jet engine or a machine, is much more complex in that the wave-form varies quite randomly as a function of time. It is not possible to say that it consists of a few discrete tones, as the sound from a musical instrument does, but it is possible to say that all frequencies are present at the same time although their relative strengths will be different. If we could select just one of those frequencies from the total random signal then the pressure would be seen to vary periodically but the amplitude would still vary randomly in time. This latter difficulty is obviated by taking the rms value measured over a reasonably long period. All of the frequencies cannot, of course, be considered in this manner, because they are infinite in number, but they can be grouped into various frequency bands and then the total strength of all the components in each band can be measured.

Octave bands are those most commonly used. The word octave originally had the connotation of halving or doubling the frequency. For convenience, however, this concept has been modified very slightly so that ten octaves correspond to a frequency ratio of 1000 and not 1024. 1000 Hz is the internationally accepted reference frequency and is the centre frequency of one octave band. The centre frequencies of other bands are obtained by continually multiplying or dividing the previous centre frequency by $10^{3/10}$ starting at 1000 Hz. The nine octave bands covering most of the audio range are listed in Table A.2 in the appendix at the end of the book. The frequency limits of each band are obtained by dividing or multiplying the centre frequency by $10^{3/20}$. So the 1000 Hz band extends from 709 to 1412 Hz. The limiting frequencies are also shown in Table A.2, in the appendix.

Sometimes octave bands are not sufficiently selective in frequency and then it is more convenient to use 1/3 octaves. This consists of dividing each octave band into three parts. The centre frequency of one 1/3 octave band will be the same as that of the octave band and the centre frequencies of the other two are obtained by multiplying or dividing the first centre frequency by $10^{1/10}$. Thus the three bands contained within the 1000 Hz octave have centre frequencies of 800, 1000 and 1250 Hz. The limiting frequencies of the 1/3 octave bands are obtained by multiplying or dividing each centre frequency by $10^{1/20}$. These are all listed in Table A.2, in the appendix.

So a noise may be analysed into frequency bands, each of which will have a band level in decibels. If these levels are plotted as a function of frequency then the resulting graph is called a spectrum. Typical spectra are illustrated in Fig. 1.4. Naturally the 1/3 octave band spectrum gives a little more information about the frequency content than the 1/1 octave band spectrum but the extra work involved is not often justified. Note that the frequency scale in Fig. 1.4 is drawn logarithmically, *i.e.* there is a constant linear distance between the octave band centre frequencies. Frequency scales are always drawn this way in acoustics.

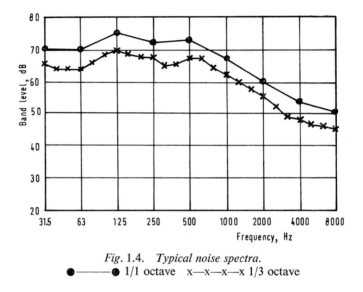

Fig. 1.4.   *Typical noise spectra.*
●————● 1/1 octave   x—x—x—x 1/3 octave

1.9   Outdoor Propagation of Sound

Sound which travels in the open can be influenced by several different factors during its path between the source and recipient. First there is the natural reduction with distance as the sound spreads out, secondly there are the effects of wind and temperature gradients, then there is the possibility of ground and air absorption and finally the sound may have to bend around some sort of obstacle which can be thought of as a barrier. All of these factors can be assessed individually and so they will be discussed one by one.

(a) *Radiation from the Source*

A small source radiating equally in all directions is the simplest which can be considered. If the acoustic power output, W, of the source is known then at a radius r the power must be equally distributed over the spherical surface area and the intensity is

$$I = W/4\pi r^2 \; Wm^{-2}$$

If the source is on the ground and can only radiate over a hemisphere then the intensity is doubled.

i.e. $$I = W/2\pi r^2 \; Wm^{-2}$$

Assuming hemispherical radiation, the intensity level in decibels may be written

$$IL = 10 \log_{10} \frac{W/2\pi r^2}{10^{-12}} \; dB$$

$$= 10 \log_{10} \frac{W/2\pi}{10^{-12}} - 20 \log_{10} r$$

The dependence upon distance is contained entirely in the last term and it is quickly seen that the intensity level reduces by 6 dB every time that the distance from the source is doubled. This is known as the inverse square law and is very useful when trying to estimate the level at one point from a knowledge of the level at some other point.

Some small sources do not radiate noise equally in all directions and then the directivity must be taken into account when calculating the level from the source power. The inverse square law may still be applied along any one radius from the source.

A line source is another possibility; examples would be long trains or steady streams of traffic. If the acoustic power output per unit length is W and it radiates over half a cylinder, since it is located on the ground, the intensity at a radius r is

$$I = W/\pi r \; Wm^{-2}$$

The intensity level is

$$IL = 10 \log_{10} \frac{W/\pi r}{10^{-12}} \; dB$$

$$= 10 \log_{10} \frac{W/\pi}{10^{-12}} - 10 \log_{10} r$$

In this case the intensity level only decreases by 3 dB for every doubling of distance.

(b) *Wind and Temperature Gradients*

In the open air over fairly flat country both wind and temperature gradients normally exist. A wind gradient results from the ground friction which reduces the wind speed close to the surface. Sound waves travel at the appropriate velocity relative to the air and so their velocity relative to the ground is the sum of the sound velocity plus the wind velocity. Downwind the total velocity increases with height above the ground and the sound waves are bent back toward the. ground giving an increase in the sound level at some distant points Upwind, however, the total velocity decreases with height above the

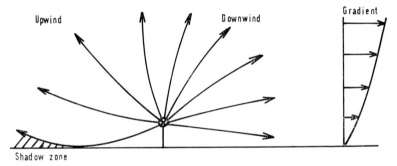

Fig. 1.5.   *Effect of a wind gradient.*

ground and the sound waves are bent upwards. This can produce a shadow zone in which, theoretically, the sound cannot be heard. In practice, diffraction and scattering from local turbulence permit some of the sound to enter the shadow zone, but there can still be an appreciable reduction. The effect of wind is illustrated in Fig. 1.5.

Temperature gradients affect the sound waves because the velocity of sound is a function of temperature. A negative gradient, such as might occur on a sunny day when the ground is warm, results in the velocity decreasing with height and the sound waves tend to bend upwards away from the ground. On a cold night, however, a positive gradient can exist and then the velocity increases with altitude so causing the sound waves to bend back towards the ground. Temperature gradients are symmetrical about the source unlike the wind gradient effect previously described. Fig. 1.6 illustrates this point.

The precise influence of both wind and temperature gradients is very difficult to predict since it depends upon the local terrain and numerous meteorological details. To give some idea of magnitude it can be said that normal wind gradients are unlikely to give an increase

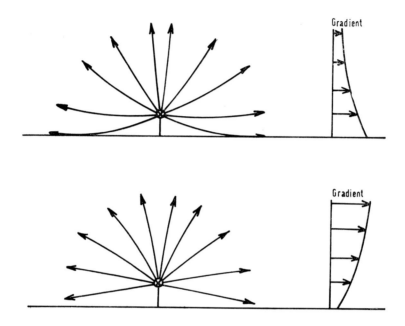

Fig. 1.6.   *Effect of temperature gradients:* (*a*) *Negative gradient,* (*b*) *Positive gradient.*

of more than 5 dB or a decrease of more than about 20 dB. Temperature gradients could give either an increase or a decrease of up to 5 dB. In the long term both wind and temperature effects tend to cancel out and so there is usually no need for an accurate calculation.

(c) *Air and Ground Absorption*

As sound travels through the air there is some molecular absorption which is a function of temperature, relative humidity and frequency. It is shown in some detail in Fig. 6.10 but we can conclude immediately that it is only significant at frequencies of 1000 Hz or more. Typical values for a temperature of 20°C and a relative humidity of 60% are shown in Table 1.2.

Ground absorption only affects those sound waves which are travelling close to the ground. Then it is a function of surface and

TABLE 1.2

AIR ABSORPTION (20°C, 60% R.H.)

Frequency in Hz	Attenuation in dB.m^{-1}
1000	0·003
2000	0·008
4000	0·025
8000	0·083

frequency. Asphalt or concrete have a negligible effect but grass can provide a useful reduction. It is only possible to give a very rough guide because there are so many variables involved but Table 1.3 gives some typical values.[1]

TABLE 1.3

ATTENUATION OF LONG GRASS AND TREES

Frequency in Hz	Attenuation in dB m^{-1}			
	Thin grass 0·1–0·2 m high	Thick grass 0·4–0·5 m high	Evergreen trees	Deciduous trees
125	0·005	0·005	0·07	0·02
250	—	—	0·11	0·04
500	—	—	0·14	0·06
1000	0·03	0·12	0·17	0·09
2000	—	—	0·19	0·12
4000	—	0·15	0·20	0·16

Trees might also be considered at this point. They give some reduction by a process of scattering and direct absorption as the sound wave travels past, but it is often disappointingly small, unless the belt of trees is extremely thick. A single line of trees, for example, has a negligible effect. The attenuation depends upon the type of tree and the density of planting and typical values are shown in Table 1.3.[2]

(d) *Effect of Barriers*

Any solid obstacle can be regarded as a barrier and the immediate point to grasp is that it does not cast a total shadow as with light. The wavelength of the sound is comparable with the dimensions of

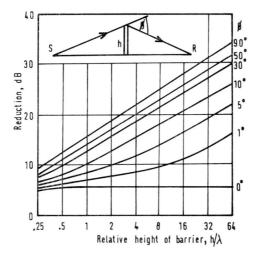

Fig. 1.7. *Reduction by a barrier.*

the barrier and diffraction of the wave around the barrier is appreciable. The larger wavelength low frequencies bend around more easily than the high frequencies and so the reduction provided by the barrier is a function of frequency, effective barrier height and the angle through which the sound wave must turn. It has been calculated theoretically using diffraction theory and to some extent checked experimentally and the results are shown in Fig. 1.7.[3]

CHAPTER 2

Sound Waves in One Dimension

2.1 The Wave Equation in One Dimension

Sound waves are simply vibrational disturbances described by their pressure, or amplitude, and frequency, which can be propagated through any medium. In the general case, with an infinite medium, the waves will travel outwards in all directions and become progressively weaker. If, however, the waves are restricted so that they can travel only in one direction, as, for example, in a thin rod or in fluid contained within a narrow tube, the situation is very much simplified. Provided that the disturbances are small and that dissipation within the medium can be neglected, the waves travel through the medium at the appropriate velocity of sound and remain unaltered in shape or size. The equation describing such disturbances is known as the wave equation in one dimension and has the following general form.

$$\frac{\partial^2 \varphi}{\partial x^2} = \frac{1}{c^2} \frac{\partial^2 \varphi}{\partial t^2} \tag{2.1}$$

where c is the velocity of propagation of the wave. In this equation φ can be taken to represent any desired quantity such as amplitude, particle velocity or pressure since, as will be shown later, they are all related to one another. It is worth noting that equation (2.1) can also be used to describe wave motion along a string, but since that is a transverse vibration, *i.e.* the displacement is normal to the direction in which the wave is travelling, it will not be considered in this book. Sound waves, remember, are longitudinal vibrations, *i.e.* the displacement is parallel with the direction in which the wave is travelling.

Equation (2.1) is solved by the method of separating variables, that

20

is by writing

$$\varphi = F_1(x)F_2(t)$$

which leads to two separate differential equations. Recombining the individual solutions gives the most general solution of equation (2.1) which is usually written

$$\varphi = g(ct-x)+h(ct+x) \qquad (2.2)$$

where g and h are arbitrary independent functions describing the waves. Consider the first half of the solution, $g(ct-x)$. At the

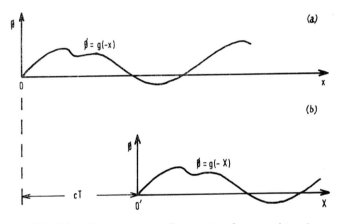

Fig. 2.1.   Representation of wave given by $\varphi = g(ct-x)$.
(a) Wave at time $t = 0$, (b) wave at time $t = T$.

particular time $t = 0$ we have $\varphi = g(-x)$ which may be represented as in Fig. 2.1 (a). At time $t = T$ we have $\varphi = g(cT-x)$ and if we change variables by writing $x = cT+X$ we have $\varphi = g(-X)$. The shape denoted by g is identical and can be represented as in Fig. 2.1 (b) where X is measured from a new origin $0'$ displaced a distance cT to the right of 0.

Evidently the wave $g(ct-x)$ has retained a constant shape but has moved a distance cT in the positive x direction at velocity c in time T. Similarly, $h(ct+x)$ represents a wave of constant shape travelling in the negative x direction, also at velocity c.

The simplest example is a harmonic wave, in which the shape is defined by a sine or a cosine, travelling in the positive x direction. This is sometimes referred to as a progressive wave. The solution may then be written

$$\varphi = a_1 \cos k(ct - x) \qquad (2.3)$$

where a_1 is the peak displacement (or velocity or pressure) and k is called the wave number.

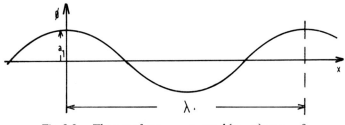

Fig. 2.2.   The waveform $\varphi = a_1 \cos k(ct - x)$ at $t = 0$.

The waveform, illustrated in Fig. 2.2, repeats itself at regular intervals in space given by the wavelength λ where

$$\lambda = \frac{2\pi}{k} \qquad (2.4)$$

Alternatively, it can be said that the waveform repeats itself at regular intervals in time given by the period T where

$$T = \frac{2\pi}{kc} = \frac{\lambda}{c} \qquad (2.5)$$

It is, of course, common practice to speak in terms of the frequency rather than the period where

$$f = \frac{1}{T} = \frac{c}{\lambda} \qquad (2.6)$$

or $$c = f\lambda \qquad (2.7)$$

Mathematically, it is preferable to use the angular frequency, $\omega = 2\pi f$ rad sec^{-1}, because equation (2.3) may then be re-written

$$\varphi = a_1 \cos (\omega t - kx) \qquad (2.8)$$

Equation (2.8) cannot completely describe any harmonic wave travelling in the positive x direction because, at any particular time, φ reaches a maximum value at particular values of x. If we choose a sine rather than a cosine then the same argument still applies but the positions at which φ reaches a maximum are different. Complete generality can be achieved by choosing a combination of sine and cosine.

i.e.
$$\varphi = a_1 \cos (\omega t - kx) + a_2 \sin (\omega t - kx) \qquad (2.9)$$

since then, by a suitable choice of constants a_1 and a_2, any desired harmonic waveform can be described.

If we are also to allow for a wave travelling in the negative x direction then this must also consist of a sine term and a cosine term and the complete harmonic solution of the wave equation must be written

$$\varphi = a_1 \cos (\omega t - kx) + a_2 \sin (\omega t - kx) + a_3 \cos (\omega t + kx)$$
$$+ a_4 \sin (\omega t + kx) \quad (2.10)$$

The waveform which really exists can be very complicated, bearing no apparent resemblance to the harmonic waves described by equation (2.10). The wave equation (2.1), however, is linear and so, as we have already seen from solution (2.2), we can write the total solution

$$\varphi = \varphi_1 + \varphi_2 + \varphi_3 + \ldots \qquad (2.11)$$

This is the principle of superposition and means that any complex waveform can be built up from a series of harmonic components. Consequently it is sufficient to consider the component parts of a complex waveform individually and the complete harmonic solution of equation (2.10) is the most general solution that need be used.

Obtaining a solution in terms of sines and cosines can become tedious and, once the basic principles are understood, it is easier to use complex algebra. The harmonic solution is written

$$\boldsymbol{\varphi} = \mathbf{A}\, e^{j(\omega t - kx)} + \mathbf{B}\, e^{j(\omega t + kx)} \qquad (2.12)$$

where $\mathbf{A}$ and $\mathbf{B}$ are both complex and it is understood that the required solution is given by the real part of the equation. To show that this is identical with equation (2.10) we can write

$$\mathbf{A} = a_1 + ja_2; \quad e^{j(\omega t - kx)} = \cos(\omega t - kx) + j\sin(\omega t - kx)$$

$$\mathbf{B} = \beta_1 + j\beta_2; \quad e^{j(\omega t + kx)} = \cos(\omega t + kx) + j\sin(\omega t + kx)$$

Then $\varphi = a_1 \cos(\omega t - kx) - a_2 \sin(\omega t - kx)$

$$+ \beta_1 \cos(\omega t + kx) - \beta_2 \sin(\omega t + kx)$$

$$+j\left[\begin{array}{l} a_1 \sin(\omega t - kx) + a_2 \cos(\omega t - kx) \\ + \beta_1 \sin(\omega t + kx) + \beta_2 \cos(\omega t + kx) \end{array}\right]$$

The real part is the same provided that $a_1 = a_1$, $a_2 = -a_2$, $\beta_1 = a_3$, $\beta_2 = -a_4$. The imaginary part of the solution is simply ignored, although it must be retained and used during the mathematical processing.

During the remainder of this chapter we shall derive particular solutions to the wave equation using both simple algebra and complex algebra. It is, however, essential that the concepts of complex algebra should be understood because in subsequent chapters only complex algebra will be used.

2.2   Longitudinal Waves in a Solid Rod

Before coming to sound waves in a gas it is useful to consider longitudinal waves in a solid. It is easier for anyone familiar with the principles of elasticity of solids to understand what is happening physically and the mathematics are quite simple.

If the rod is sufficiently thin compared with the wavelength of the disturbance considered, it may be assumed that the displacement at any instant is the same at all points on a cross-section of the rod.

Fig. 2.3 shows a rod of cross-sectional area S and density ρ_0. PQ is a small element of length δx at a distance x from the origin. At some particular instant a small disturbance, due to the passage of a longitudinal wave, causes the element to move to $P'Q'$ increasing its distance from the origin by ξ and its length by $\delta \xi$.

The tension at P' is given by Hooke's Law

i.e. $T_{P'} = E \times S \times \text{strain}$

$$= ES \frac{\partial \xi}{\partial x} \tag{2.13}$$

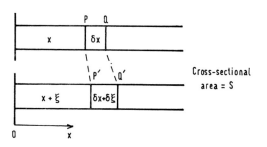

Fig. 2.3.   Displacement of an element of a solid rod due to the passage of a longitudinal wave.

The tension at Q' is equal to that at P' plus an elemental increase dependent upon the rate of change of T with respect to x

i.e. $T_{Q'} = T_{P'} + \frac{\partial T}{\partial x} \delta x \tag{2.14}$

The mass of the element $= S \rho_0 \, \delta x \tag{2.15}$

and the acceleration $= \frac{\partial^2 \xi}{\partial t^2} \tag{2.16}$

Combining equations (2.14), (2.15) and (2.16) according to Newton's law of motion gives

$$S \rho_0 \, \delta x \frac{\partial^2 \xi}{\partial t^2} = T_{Q'} - T_{P'} = \frac{\partial T}{\partial x} \delta x = ES \frac{\partial^2 \xi}{\partial x^2} \delta x$$

i.e. $\frac{\partial^2 \xi}{\partial x^2} = \frac{1}{c^2} \frac{\partial^2 \xi}{\partial t^2}$ where $c^2 = E/\rho_0 \tag{2.17}$

This is the wave equation in one dimension, which was discussed in Section 2.1, written in terms of the longitudinal displacement and the general solution is given either by equation (2.10) or (2.12).

If the rod is finite in length, or indeed if the movement of an infinite rod is restricted in some way, the general solution has to be made to fit the boundary conditions and becomes a particular solution. The common boundary conditions are:

(1) *Clamped end.* At the particular value of x where the rod is clamped the displacement is zero at all times.

i.e. $$(\xi)_x = 0 \tag{2.18}$$

(2) *Free end.* At the end of the rod, given by some particular value of x, the tension must at all times be zero. From equation (2.13) we see that the tension is proportional to $\partial \xi / \partial x$

therefore $$\left(\frac{\partial \xi}{\partial x}\right)_x = 0 \tag{2.19}$$

As an example, consider a rod of length L, free at both ends. Using simple algebra and writing the solution in the form of equation (2.10) we have

$$\xi = a_1 \cos (\omega t - kx) + a_2 \sin (\omega t - kx) + a_3 \cos (\omega t + kx)$$
$$+ a_4 \sin (\omega t + kx)$$

Then

$$\partial \xi / \partial x = a_1 k \sin (\omega t - kx) - a_2 k \cos (\omega t - kx)$$
$$- a_3 k \sin (\omega t + kx) + a_4 k \cos (\omega t + kx)$$

Applying the first boundary condition,

i.e. $$\left(\frac{\partial \xi}{\partial x}\right)_{x=0} = 0 \quad \text{at all values of } t$$

we have

$$0 = a_1 k \sin \omega t - a_2 k \cos \omega t - a_3 k \sin \omega t + a_4 k \cos \omega t$$

therefore $$a_3 = a_1 \quad \text{and} \quad a_4 = a_2$$

and the solution may be re-written

$$\xi = a_1 \{\cos (\omega t - kx) + \cos (\omega t + kx)\} + a_2 \{\sin (\omega t - kx) + \sin (\omega t + kx)\}$$
$$= 2a_1 \cos \omega t \cos kx + 2a_2 \sin \omega t \cos kx$$
$$= 2 \cos kx (a_1 \cos \omega t + a_2 \sin \omega t) \tag{2.20}$$

Equation (2.20) describes a wave in which the space and time variations have become separated. It is a stationary wave and is the sum of two equal and opposite progressive waves. In this case it may be thought of as a wave travelling in the negative x direction which is perfectly reflected at the free end located at the origin.

Fig. 2.4 illustrates the envelope of the waveform described by equation (2.20) for some particular value of k, and therefore of ω. At certain points, given by $\cos kx = 0$, the displacement ξ is always zero, and these are referred to as displacement nodes. At other points in between the nodes, given by $\cos kx = \pm 1$, the amplitude is a maximum and these positions are referred to as displacement antinodes.

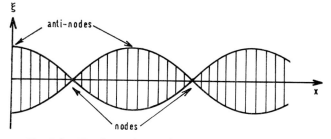

Fig. 2.4. Envelope of wave described by equation (2.20).

Differentiating equation (2.20) gives

$$\frac{\partial \xi}{\partial x} = -2k \sin kx (a_1 \cos \omega t + a_2 \sin \omega t)$$

Applying the second boundary condition,

i.e.
$$\left(\frac{\partial \xi}{\partial x}\right)_{x=L} = 0 \quad \text{at all values of } t$$

results in

$$\sin kL = 0 \tag{2.21}$$

i.e.
$$kL = n\pi \quad \text{where} \quad n \text{ is an integer}$$

Referring to equations (2.4) and (2.6) gives

$$\lambda = \frac{2L}{n}$$

and
$$f = \frac{nc}{2L} \tag{2.22}$$

So we see that, with a rod free at both ends, vibrations can only be sustained when the length of the rod contains an integral number of half-wavelengths. The velocity at which the waves travel depends on the substance of the rod and the frequencies of the vibrations depend

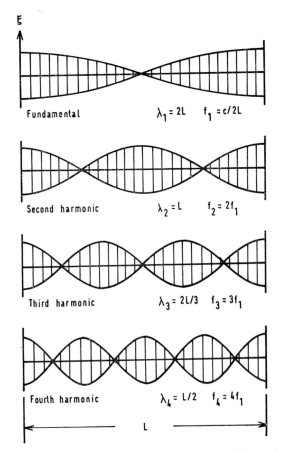

Fundamental $\lambda_1 = 2L$ $f_1 = c/2L$

Second harmonic $\lambda_2 = L$ $f_2 = 2f_1$

Third harmonic $\lambda_3 = 2L/3$ $f_3 = 3f_1$

Fourth harmonic $\lambda_4 = L/2$ $f_4 = 4f_1$

L

Fig. 2.5.   First four resonant modes of vibration of a rod free at both ends.

upon the velocity and the length of the rod. These particular vibrations are often referred to as natural or resonant vibrations and the first four resonant modes are illustrated in Fig. 2.5. The lowest resonant frequency is referred to as the fundamental and the other resonant frequencies are harmonics since they are integrally related to the fundamental.

The example could equally well have been worked out using complex algebra starting with a solution of the form given in equation (2.12).

i.e.
$$\xi = \mathbf{A}\, e^{j(\omega t - kx)} + \mathbf{B}\, e^{j(\omega t + kx)}$$

$$\frac{\partial \xi}{\partial x} = -jk\mathbf{A}\, e^{j(\omega t - kx)} + jk\mathbf{B}\, e^{j(\omega t + kx)}$$

Now
$$\left(\frac{\partial \xi}{\partial x}\right)_{x=0} = 0 \quad \text{at all values of } t$$

therefore
$$\mathbf{B} = \mathbf{A}$$

and the solution becomes

$$\xi = \mathbf{A}(e^{jkx} + e^{-jkx})\, e^{j\omega t}$$
$$= 2\mathbf{A}\cos kx\, e^{j\omega t}$$

This is directly comparable with equation (2.19) and represents a standing wave.

Differentiating with respect to x gives

$$\frac{\partial \xi}{\partial x} = -2k\mathbf{A}\sin kx\, e^{j\omega t}$$

Now
$$\left(\frac{\partial \xi}{\partial x}\right)_{x=L} = 0 \quad \text{at all values of } t$$

therefore
$$\sin kL = 0$$

This is the same result as was obtained previously showing that the particular mathematical method which is chosen has no effect on the solution which is obtained.

2.3 Longitudinal Waves in a Column of Gas

It is now time to derive the equation of motion for longitudinal waves in a gas. The general method is the same as that for a solid rod but is a little more complicated because the gas is compressible.

As with the rod, the tube should be sufficiently narrow for the displacement at any instant to be the same at all points on any cross-section. Friction along the sides of the tube is assumed not to exist.

Fig. 2.6 shows a tube of cross-sectional area S. At some instant, the element PQ is moved to $P'Q'$ by the passage of a longitudinal wave, so increasing its distance from the origin by ξ and its volume by $S\delta\xi$.

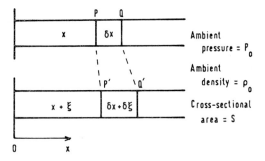

Fig. 2.6. Displacement of an element of gas in a tube due to the passage of a longitudinal wave.

The principles involved in obtaining the equation are based on Newton's law of motion and the elastic behaviour of the gas.

(1) *Newton's law of motion*

The net force acting on the displaced element

$$= (P_{P'} - P_{Q'})S$$
$$= -\frac{\partial P}{\partial x} S\,\delta x$$

It is convenient to write the total pressure P in terms of a small pressure variation, p, superimposed on the ambient pressure P_0, since P_0 then disappears from any differential leaving only the pressure variation. The net force on the element may be written

$$\text{Net force} = -\frac{\partial p}{\partial x} S\,\delta x \qquad (2.23)$$

Now the mass of element $P'Q'$ = mass of element PQ

$$= S\rho_0 \, \delta x \qquad (2.24)$$

and the acceleration of the element

$$= \frac{\partial^2 \xi}{\partial t^2} \qquad (2.25)$$

Combining these three equations gives

$$\frac{\partial p}{\partial x} = -\rho_0 \frac{\partial^2 \xi}{\partial t^2} \qquad (2.26)$$

(2) *Elastic behaviour*

Since the gas is compressible a change of pressure gives rise to a significant change of volume. Normally the pressure fluctuates quite rapidly, at least 20 Hz, and it can be assumed that the process is adiabatic. Therefore we may write

$$P_0 V_0^\gamma = K$$

where V_0 is the undisturbed volume, γ is the ratio of the specific heats of the gas and K is a constant.

The differential may be written

$$\gamma P_0 V_0^{\gamma-1} \, \delta V + V_0^\gamma \, \delta P = 0$$

δP is the same as the small pressure variation, p, and this equation becomes

$$p = -\gamma P_0 \frac{\delta V}{V_0} \qquad (2.27)$$

In this case the undisturbed volume, V_0, is $S \, \delta x$ and the change in volume is $S \, \delta \xi$.

therefore $$p = -\gamma P_0 \frac{\partial \xi}{\partial x} \qquad (2.28)$$

This equation may be differentiated with respect to x and combined with equation (2.26) to eliminate p giving

$$\gamma P_0 \frac{\partial^2 \xi}{\partial x^2} = \rho_0 \frac{\partial^2 \xi}{\partial t^2}$$

i.e. $$\frac{\partial^2 \xi}{\partial x^2} = \frac{1}{c^2} \frac{\partial^2 \xi}{\partial t^2} \qquad (2.29)$$

where $$c^2 = \gamma P_0/\rho_0 \qquad (2.30)$$

So we have obtained the wave equation in one dimension in terms of the displacement ξ. In acoustics, however, we are often more interested in the small pressure variation, p, and this is given in terms of the displacement by equation (2.28). The substitution of $\rho_0 c^2$ for γP_0 is commonly made, so that

$$p = -\rho_0 c^2 \frac{\partial \xi}{\partial x} \qquad (2.31)$$

It is also possible to obtain the wave equation directly in terms of pressure by differentiating equation (2.26) with respect to x.

$$\frac{\partial^2 p}{\partial x^2} = -\rho_0 \frac{\partial^2}{\partial t^2} \left(\frac{\partial \xi}{\partial x} \right)$$

Substitution for $\partial \xi/\partial x$ from equation (2.31) yields

$$\frac{\partial^2 p}{\partial x^2} = \frac{1}{c^2} \frac{\partial^2 p}{\partial t^2} \qquad (2.32)$$

We shall in future use this equation in preference to equation (2.29) so that we can write the solution directly in terms of pressure and also because the wave equation in three dimensions, which will be discussed in the next chapter, can be formulated most easily in terms of pressure. The equations in displacement and pressure only have exactly the same form in this simple one-dimensional case.

The other quantity in which we shall be interested is the particle velocity, u, and this can be obtained in terms of pressure by writing equation (2.26) in the form

$$\frac{\partial u}{\partial t} = -\frac{1}{\rho_0} \frac{\partial p}{\partial x}$$

Integration yields

$$u = -\frac{1}{\rho_0} \int \frac{\partial p}{\partial x} \partial t \qquad (2.33)$$

The particle velocity associated with a one-dimensional simple harmonic wave travelling in the positive x direction can be calculated once and for all

If $$p_+ = a_1 \cos(\omega t - kx) + a_2 \sin(\omega t - kx)$$

then $$\frac{\partial p_+}{\partial x} = ka_1 \sin(\omega t - kx) - ka_2 \cos(\omega t - kx)$$

$$\int \frac{\partial p_+}{\partial x} \partial t = -\frac{k}{\omega} a_1 \cos(\omega t - kx) - \frac{k}{\omega} a_2 \sin(\omega t - kx)$$

$$= -\frac{k}{\omega} p_+$$

and $$u_+ = \frac{k}{\omega \rho_0} p_+ = \frac{p_+}{\rho_0 c} \qquad (2.34)$$

If we considered a simple harmonic wave travelling in the negative x direction then we would find that

$$u_- = -\frac{p_-}{\rho_0 c}$$

These relationships between pressure and particle velocity are very useful since one-dimensional wave systems can always be split into two components, one travelling in each direction, and then the particle velocity of each is obtained directly from the pressure of each by dividing by $\pm \rho_0 c$, depending upon the direction of the wave. The total pressure and particle velocity are, of course, given by

$$p = p_+ + p_- \quad \text{and} \quad u = u_+ + u_- \qquad (2.35)$$

When the tube is of finite length boundary conditions will have to be satisfied, so reducing the general solutions to particular ones. Two common boundary conditions are

(1) *Closed tube.* At the particular value of x at which the tube is closed, the velocity must at all times be zero.

i.e. $$(u)_x = 0 \qquad (2.36)$$

(2) *Open tube.* If the wave did not extend beyond the tube the pressure fluctuation would at all times be zero.

i.e. $$(p)_x = 0 \qquad (2.37)$$

In fact, because the fluid is the same both inside and outside the tube, the disturbances extend slightly beyond the open end. It is therefore usual to suppose that the tube is slightly extended and to apply the boundary condition at the imaginary end. The end correction has been shown to be 0·6a for a tube of radius a with no flange and $8a/3\pi$ for a tube with a large flange.

As an example consider an unflanged tube of length L, closed at one end and open at the other end. Writing the pressure solution in terms of sines and cosines

$$p = a_1 \cos (\omega t - kx) + a_2 \sin (\omega t - kx) + a_3 \cos (\omega t + kx)$$
$$+ a_4 \sin (\omega t + kx)$$

The particle velocity is

$$u = \frac{a_1}{\rho_0 c} \cos (\omega t - kx) + \frac{a_2}{\rho_0 c} \sin (\omega t - kx) - \frac{a_3}{\rho_0 c} \cos (\omega t + kx)$$
$$- \frac{a_4}{\rho_0 c} \sin (\omega t + kx)$$

Applying the first boundary condition

i.e. $(u)_{x=0}$ at all values of t

then $0 = a_1 \cos \omega t + a_2 \sin \omega t - a_3 \cos \omega t - a_4 \sin \omega t$

therefore $a_3 = a_1; \quad a_4 = a_2$

and the solution becomes

$$p = a_1 \{\cos (\omega t - kx) + \cos (\omega t + kx)\} + a_2 \{\sin (\omega t - kx) + \sin (\omega t + kx)\}$$
$$= 2a_1 \cos \omega t \cos kx + 2a_2 \sin \omega t \cos kx$$
$$= 2 \cos kx (a_1 \cos \omega t + a_2 \sin \omega t) \tag{2.38}$$

The waveform is stationary and a pressure maximum, or antinode, occurs at the rigid end located at the origin.

The second boundary condition, namely that the tube is open at $x = L$, may now be applied. It must be remembered that there is an end correction and the actual condition is that

$$(p)_{x = L'} = 0 \quad \text{at all values of } t$$

where $L' = L + 0·6a$

therefore $\cos kL' = 0 \tag{2.39}$

i.e. $kL' = (2n - 1)\pi/2$

where n is an integer.

Referring to equations (2.4) and (2.6) gives

$$\lambda = \frac{4L'}{(2n-1)} \qquad (2.40)$$

and

$$f = \frac{(2n-1)c}{4L'} \qquad (2.41)$$

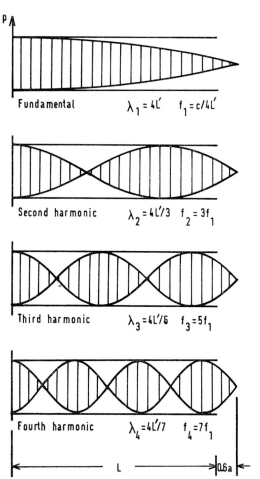

Fundamental $\qquad \lambda_1 = 4L' \qquad f_1 = c/4L'$

Second harmonic $\qquad \lambda_2 = 4L'/3 \qquad f_2 = 3f_1$

Third harmonic $\qquad \lambda_3 = 4L'/5 \qquad f_3 = 5f_1$

Fourth harmonic $\qquad \lambda_4 = 4L'/7 \qquad f_4 = 7f_1$

L $\qquad$ 0.6a

Fig. 2.7. Pressure amplitudes associated with the first four modes of vibration of an unflanged pipe of length L closed at one end and open at the other.

As with the rod considered in the previous section, the pipe will only sustain natural vibrations or resonances at certain frequencies. The velocity of sound depends upon the gas and the resonant frequencies are a function of the length of the tube and of the velocity. The lowest resonant frequency is referred to as the fundamental, and the higher resonant frequencies are harmonics. In this case, because of the particular boundary conditions which are imposed, the frequencies of the harmonics are all odd integral multiples of the fundamental frequency.

The example could just as easily have been worked out using complex algebra. We assume a general solution of the form.

$$\mathbf{p} = \mathbf{A}\, e^{j(\omega t - kx)} + \mathbf{B}\, e^{j(\omega t + kx)}$$

therefore
$$\mathbf{u} = \frac{\mathbf{A}}{\rho_0 c}\, e^{j(\omega t - kx)} - \frac{\mathbf{B}}{\rho_0 c}\, e^{j(\omega t + kx)}$$

Now
$$(\mathbf{u})_{x=0} = 0 \quad \text{for all values of } t$$

therefore
$$\mathbf{B} = \mathbf{A}$$

and the solution becomes

$$\mathbf{p} = \mathbf{A}(e^{jkx} + e^{-jkx})\, e^{j\omega t}$$
$$= 2\mathbf{A} \cos kx\, e^{j\omega t}$$

If the pressure is to be zero at $x = L'$ for all values of t

then
$$\cos kL' = 0$$

which is the same as equation (2.39).

2.4   Energy Density and Intensity

So far we have only discussed sound waves in terms of displacement, particle velocity and pressure. The pressure, particularly, is a useful and meaningful quantity since it is that which actuates microphones and, indeed, our ears. But, in addition, we are interested in energy, since it is a more fundamental quantity, and so we must find some relationship between pressure and energy.

When a sound wave is transmitted we already know that the gas particles vibrate and that the gas undergoes a series of adiabatic compressions and rarefactions. Therefore, the energy density, which is the energy contained within unit volume of the fluid and may be a function of both space and time, must consist of two parts

1. The kinetic energy of the particles.
2. The potential energy of the gas.

The kinetic energy is easily obtained, provided that the particle velocity, u, is known, and is given by

$$D_{KE} = \tfrac{1}{2}\rho_0 u^2 \qquad (2.42)$$

The potential energy gained by unit volume of a gas of volume V_0 when it undergoes an expansion dV is, by definition

$$D_{PE} = -\frac{1}{V_0} \int p\,dV \qquad (2.43)$$

Differentiating the adiabatic law

$$PV^\gamma = K$$

gives

$$\frac{dV}{V_0} = -\frac{1}{\gamma}\frac{dp}{P_0} \qquad (2.44)$$

Substituting equation (2.44) into equation (2.43) gives, for the potential energy

$$D_{PE} = \frac{1}{\gamma P_0} \int p\,dp$$

$$= \frac{1}{2}\frac{p^2}{\gamma P_0}$$

$$= \frac{1}{2}\frac{p^2}{\rho_0 c^2} \qquad (2.45)$$

So the total energy density at any given instant of time is

$$D(x,t) = \frac{1}{2}\left(\rho_0 u^2 + \frac{p^2}{\rho_0 c^2}\right) \qquad (2.46)$$

The sound intensity describes the flow rate of sound energy through unit area of the gas. It may be a function of both time and space and it is a vector quantity since the energy is flowing in a particular direction. The intensity, then, is the rate at which the pressure does work and it is equal to the product of the pressure and that component of the particle velocity which is in phase with the pressure.

i.e. $$I(x,t) = pu \qquad (2.47)$$

where it is assumed that the expressions for p and u contain phase information.

Let us first calculate the energy density and intensity of a plane progressive wave travelling in the positive x direction. It is sufficient in this case to describe the wave by taking only the first term of equation (2.10)

i.e. $$p = a_1 \cos{(\omega t - kx)}$$

and the particle velocity

$$u = \frac{p}{\rho_0 c}$$

Substitution into equation (2.46) gives the energy density

$$D(x,t) = \frac{1}{2} \left(\frac{p^2}{\rho_0 c^2} + \frac{p^2}{\rho_0 c^2} \right)$$

$$= \frac{a_1^2}{\rho_0 c^2} \cos^2{(\omega t - kx)}$$

At any point the energy density fluctuates with time as the pressure fluctuates. An average value can be obtained by integrating over one period and dividing by that period. Furthermore, since we are dealing with a plane progressive wave it follows that the average energy density must be the same at all points in space.

therefore $$D = \frac{a_1^2}{\rho_0 c^2} \frac{1}{T} \int_0^T \cos^2{(\omega t - kx)} \, \partial t$$

$$= \frac{a_1^2}{2 \rho_0 c^2} \qquad (2.48)$$

Any pressure sensing device will be unable to measure a_1 because it will tend to average the pressure with respect to time. It will, in fact, measure the root mean square pressure, $\bar{p}$, which can be obtained in terms of the peak pressure by the same simple integral with respect to time as was performed above which gives

$$\bar{p}^2 = \frac{1}{T} \int_0^T a_1{}^2 \cos^2 (\omega t - kx) \, \partial t$$

$$= \frac{a_1{}^2}{2} \tag{2.49}$$

Therefore, the average energy density in terms of the rms pressure is given by

$$D = \frac{\bar{p}^2}{\rho_0 c^2} \tag{2.50}$$

The intensity is obtained by substituting the expressions for pressure and particle velocity into equation (2.47) to give

$$I(x,t) = \frac{a_1{}^2}{\rho_0 c} \cos^2 (\omega t - kx)$$

As it happens pressure and particle velocity are exactly in phase. Averaging in time as we did before and remembering that the average intensity must be the same for all values of x, we obtain for the average intensity

$$I = \frac{a_1{}^2}{2\rho_0 c} \tag{2.51}$$

$$= \frac{\bar{p}^2}{\rho_0 c} \tag{2.52}$$

Comparing equations (2.50 and 2.52) we see that

$$I = Dc \tag{2.53}$$

This is a result which we might have anticipated since the intensity is the quantity of energy which passes through unit cross-sectional area in unit time and must be equal to the total quantity of energy contained within a column of unit cross-sectional area and of a length numerically equal to the velocity of sound.

It is worth using complex algebra on the same example since it yields an important result which we shall use in subsequent chapters.

The expressions for the wave are

$$\mathbf{p} = \mathbf{A}\, e^{j(\omega t - kx)}$$

and

$$\mathbf{u} = \frac{\mathbf{A}}{\rho_0 c}\, e^{j(\omega t - kx)}$$

The actual pressure and particle velocity are given by the real parts of these expressions and so it is necessary to multiply together only

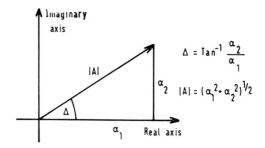

$$\Delta = \mathrm{Tan}^{-1}\,\frac{\alpha_2}{\alpha_1}$$

$$|A| = (\alpha_1{}^2 + \alpha_2{}^2)^{1/2}$$

Fig. 2.8. *Representation of the vector* **A** *in the complex plane.*

the real parts for calculating energy. To do this it is easiest to write

$$\mathbf{A} = |\mathbf{A}|\, e^{j\Delta}$$

The relationships between $|\mathbf{A}|$, Δ and α_1, α_2 are shown in Fig. 2.8. Then, taking the real parts only

$$p = |\mathbf{A}|\cos(\Delta + \omega t - kx)$$

$$u = \frac{|\mathbf{A}|}{\rho_0 c}\cos(\Delta + \omega t - kx)$$

The energy density

$$D(x,t) = \frac{|\mathbf{A}|^2}{\rho_0 c^2}\cos^2(\Delta + \omega t - kx)$$

Averaging in time and noting that there is no spatial variation gives

$$D = \frac{|\mathbf{A}|^2}{2\rho_0 c^2} \qquad (2.54)$$

The mean square pressure

$$\bar{p}^2 = \frac{|A|^2}{2} \tag{2.55}$$

and $D = \bar{p}^2/\rho_0 c^2$ as before.
Similarly the intensity

$$I = \frac{|A|^2}{2\rho_0 c}$$

$$= \frac{\bar{p}^2}{\rho_0 c} \tag{2.56}$$

For a wave which is expressed in the complex form we conclude that it is the moduli (corresponding to the peak values) of the pressure and the particle velocity which are important for energy considerations. Generally the energy density consists of two terms, one containing p^2 and the other containing u^2 so the phase between the pressure and the particle velocity is unimportant. The expression for the average energy density in terms of moduli is given by

$$D(x) = \frac{1}{2} \left(\rho_0 \frac{|\mathbf{u}|^2}{2} + \frac{|\mathbf{p}|^2}{2\rho_0 c^2} \right) \tag{2.57}$$

For the intensity, however, we need to know the phase difference θ between pressure and particle velocity so that we can write

$$I(x) = \frac{|\mathbf{p}| \times |\mathbf{u}|}{2} \cos \theta \tag{2.58}$$

It is also of interest to calculate the energy density and intensity in the other extreme case of a one-dimensional standing wave. Using simple algebra, it is sufficient to write the pressure in the form

$$p = a_1 \cos kx \cos \omega t \tag{2.59}$$

therefore

$$u = \frac{a_1}{\rho_0 c} \sin kx \sin \omega t \tag{2.60}$$

The energy density

$$D(x,t) = \frac{1}{2} \left(\frac{a_1^2}{\rho_0 c^2} \sin^2 kx \sin^2 \omega t + \frac{a_1^2}{\rho_0 c^2} \cos^2 kx \cos^2 \omega t \right)$$

$$= \frac{a_1^2}{2\rho_0 c^2} (\sin^2 kx \sin^2 \omega t + \cos^2 kx \cos^2 \omega t)$$

Averaging first in time gives

$$D(x) = \frac{a_1^2}{4\rho_0 c^2} (\sin^2 kx + \cos^2 kx) \qquad (2.61)$$

Always appears as   Always appears as
kinetic energy   potential energy

and therefore

$$D = \frac{a_1^2}{4\rho_0 c^2} \qquad (2.62)$$

Again the average energy density is the same at all points in space, although in this case it always appears as kinetic energy at the pressure nodes or velocity antinodes and always appears as potential energy at the pressure antinodes or velocity nodes. Equation (2.62) gives the energy density in terms of the peak pressure which only ever occurs at the pressure antinodes. First of all it is necessary to average this pressure in time which introduces a factor of two. Secondly it is necessary to average in space which introduces a second factor of two so that the mean rms pressure $\bar{p}$ is given by

$$\bar{p}^2 = \frac{a_1^2}{4} \qquad (2.63)$$

therefore

$$D = \frac{\bar{p}^2}{\rho_0 c^2} \qquad (2.64)$$

which is identical to equation (2.50) although it has been necessary to average the pressure in both time and space to achieve this result.

The intensity is obtained by multiplying together the pressure and the particle velocity

$$I(x,t) = \frac{a_1^2}{\rho_0 c} \sin kx \cos kx \sin \omega t \cos \omega t$$

and if we average in time we get

$$I(x) = 0 \qquad (2.65)$$

This result should have been expected since, by definition, energy is not being transmitted through a standing wave.

Alternatively we could suppose that the standing wave of equation (2.59) comprises two equal and opposite travelling waves

i.e.
$$p = \frac{a_1}{2} \cos(\omega t - kx) + \frac{a_1}{2} \cos(\omega t + kx)$$

Then
$$u = \frac{a_1}{2\rho_0 c} \cos(\omega t - kx) - \frac{a_1}{2\rho_0 c} \cos(\omega t + kx)$$

The energy density and intensity associated with each travelling wave can be calculated separately and, by referring to equations (2.48) and (2.51), we obtain

$$D_+ = \frac{a_1^2}{8\rho_0 c^2} \qquad D_- = \frac{a_1^2}{8\rho_0 c^2}$$

$$I_+ = \frac{a_1^2}{8\rho_0 c} \qquad I_- = -\frac{a_1^2}{8\rho_0 c}$$

So the total energy density and intensity are given by

$$D = D_+ + D_- = \frac{a_1^2}{4\rho_0 c^2}$$

and
$$I = I_+ + I_- = 0$$

which confirm equations (2.62) and (2.65).
The intensity in just one direction

$$I_+ = \frac{Dc}{2} = \frac{\bar{p}^2}{2\rho_0 c} \tag{2.66}$$

If we are to do the same example using complex algebra we must write

$$\mathbf{p} = \frac{\mathbf{A}}{2} e^{j(\omega t - kx)} + \frac{\mathbf{A}}{2} e^{j(\omega t + kx)}$$

$$= \mathbf{A} \cos kx \, e^{j\omega t}$$

$$= |\mathbf{A}| \cos kx \, e^{j(\Delta + \omega t)}$$

and
$$\mathbf{u} = -j \frac{|\mathbf{A}|}{\rho_0 c} \sin kx \, e^{j(\Delta + \omega t)}$$

The j in the expression for the particle velocity denotes that there is a $\pi/2$ phase difference between velocity and pressure. Using equations (2.57) and (2.58)

$$D(x) = \frac{1}{2} \left(\rho_0 \frac{|\mathbf{A}|^2}{2\rho_0^2 c^2} \sin^2 kx + \frac{|\mathbf{A}|^2 \cos^2 kx}{2\rho_0 c^2} \right)$$

$$= \frac{|\mathbf{A}|^2}{4\rho_0 c^2} (\sin^2 kx + \cos^2 kx)$$

therefore $$D = \frac{|\mathbf{A}|^2}{4\rho_0 c^2} \qquad (2.67)$$

The mean square pressure, averaged in both time and space, is

$$\bar{p}^2 = \frac{|\mathbf{A}|^2}{4} \qquad (2.68)$$

and so $$D = \frac{\bar{p}^2}{\rho_0 c^2}$$

The intensity

$$I(x) = \frac{|\mathbf{A}| \cos kx \times |\mathbf{A}| \sin kx}{2\rho_0 c} \cos \frac{\pi}{2}$$

$$= 0$$

Alternatively we could think in terms of the two equal and opposite travelling waves, and make use of equations (2.54) and (2.56).

Then $$D_+ = \frac{|\mathbf{A}|^2}{8\rho_0 c^2} \qquad D_- = \frac{|\mathbf{A}|^2}{8\rho_0 c^2}$$

$$I_+ = \frac{|\mathbf{A}|^2}{8\rho_0 c} \qquad I_- = -\frac{|\mathbf{A}|^2}{8\rho_0 c}$$

CHAPTER 3

Sound Waves in Three Dimensions

3.1 The Wave Equation in Three Dimensions

The use of the one-dimensional wave equation, developed in the previous chapter, is clearly limited since most acoustic problems are in three dimensions. The general equation describing the propagation of a wave through a gas is, however, developed along the same general lines as the one-dimensional equation. Either Cartesian or spherical coordinates could be used, but it is easier to see the development from the one-dimensional case with Cartesian coordinates. The final equation is independent of the coordinates used in the derivation and the coordinates chosen for the subsequent solution of the equation are suited to the particular case. Clearly spherical coordinates would be used for spherical radiation, cylindrical coordinates for cylindrical radiation and Cartesian coordinates for sound waves in a rectangular enclosure.

The equilibrium position of a particle of fluid has the three coordinates x, y and z. A displacement of this particle has components ξ, η and ζ and the particle velocity has components $\partial\xi/\partial t$, $\partial\eta/\partial t$ and $\partial\zeta/\partial t$.

To set up the wave equation we must make use of the same fundamental principles as we used in Chapter 2.

(1) Newton's law of motion

Motion in the three directions is unrelated since the acceleration in a direction is proportional to the force in that direction. So equation (2.26) is written in terms of each of the three coordinates x, y and z.

i.e. $$\frac{\partial p}{\partial x} = -\rho_0 \frac{\partial^2 \xi}{\partial t^2}; \quad \frac{\partial p}{\partial y} = -\rho_0 \frac{\partial^2 \eta}{\partial t^2}; \quad \frac{\partial p}{\partial z} = -\rho_0 \frac{\partial^2 \zeta}{\partial t^2} \qquad (3.1)$$

These are combined by differentiating each with respect to its own coordinate and adding

i.e. $$\left(\frac{\partial^2 p}{\partial x^2} + \frac{\partial^2 p}{\partial y^2} + \frac{\partial^2 p}{\partial z^2}\right) = -\rho_0 \frac{\partial^2}{\partial t^2}\left(\frac{\partial \xi}{\partial x} + \frac{\partial \eta}{\partial y} + \frac{\partial \zeta}{\partial z}\right) \qquad (3.2)$$

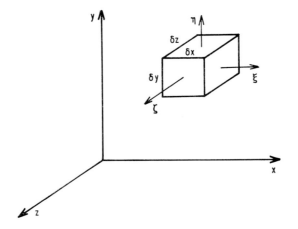

Fig. 3.1.   *Element of gas in three dimensions.*

(2) *Elastic behaviour*

This was developed for the general case in Chapter 2 and equation (2.27) remains unaltered.

i.e. $$p = -\gamma P_0 \frac{\delta V}{V_0} \qquad (3.3)$$

If we suppose that the passage of a sound wave extends the three sides of the volume element shown in Fig. 3.1 by the amounts $\delta\xi$, $\delta\eta$ and $\delta\zeta$ then the increase in volume is approximately

$$\delta V = \delta\zeta\delta y\delta z + \delta\eta\delta x\delta z + \delta\zeta\delta x\delta y$$

and, since
$$V_0 = \delta x \delta y \delta z$$

$$\frac{\delta V}{V_0} = \frac{\partial \zeta}{\partial x} + \frac{\partial \eta}{\partial y} + \frac{\partial \zeta}{\partial z}$$

therefore
$$p = -\gamma P_0 \left(\frac{\partial \zeta}{\partial x} + \frac{\partial \eta}{\partial y} + \frac{\partial \zeta}{\partial z} \right) \qquad (3.4)$$

Substitution of equation (3.4) into equation (3.2) gives

$$\left(\frac{\partial^2 p}{\partial x^2} + \frac{\partial^2 p}{\partial y^2} + \frac{\partial^2 p}{\partial z^2} \right) = \frac{1}{c^2} \frac{d^2 p}{\partial t^2} \qquad (3.5)$$

or
$$\nabla^2 p = \frac{1}{c^2} \frac{\partial^2 p}{\partial t^2} \qquad (3.6)$$

This is the general wave equation and clearly reduces to the one-dimensional form if there is no variation of pressure in the y or z directions.

The particle velocity is related to the pressure through equation (3.1).

$$u = \frac{d\xi}{\partial t} + \frac{\partial \eta}{\partial t} + \frac{\partial \zeta}{\partial t}$$

$$= -\frac{1}{\rho_0} \int \left(\frac{\partial p}{\partial x} + \frac{\partial p}{\partial y} + \frac{\partial p}{\partial z} \right) \partial t$$

or
$$u = -\frac{1}{\rho_0} \int \text{grad} \, (p) \, \partial t \qquad (3.7)$$

If it is necessary to solve the equation in some other coordinate system, we require definitions of ∇^2 and grad in that system. We shall be considering spherical radiation and, in the simplest case of spherical symmetry, it may be shown that

$$\nabla^2 = \left(\frac{\partial^2}{\partial r^2} + \frac{2}{r} \frac{\partial}{\partial r} \right)$$

$$\text{grad} = \frac{\partial}{\partial r}$$

So, in spherical coordinates

$$\left(\frac{\partial^2 p}{\partial r^2} + \frac{2}{r}\frac{\partial p}{\partial r}\right) = \frac{1}{c^2}\frac{\partial^2 p}{\partial t^2}$$

which can be rewritten

$$\frac{\partial^2(rp)}{\partial r^2} = \frac{1}{c^2}\frac{\partial^2(rp)}{\partial t^2} \tag{3.8}$$

and

$$u = -\frac{1}{\rho_0}\int \frac{\partial p}{\partial r}\,\partial t \tag{3.9}$$

3.2   Spherical Waves in Free Space

If we assume that the sound field is created by a small spherically symmetric source and that there are no boundaries to reflect any of the sound waves, equations (3.8) and (3.9) are the most appropriate to use. By analogy with equations (2.1) and (2.2) the general solution is

$$pr = g(ct-r) + h(ct+r)$$

or

$$p = \frac{1}{r}g(ct-r) + \frac{1}{r}h(ct+r) \tag{3.10}$$

The first part of the solution is an outgoing wave starting from the origin and the second part is an incoming wave converging on the origin. The terms $1/r$ imply that the pressure decreases as r increases, which is quite reasonable. That the solution becomes infinite when r is zero is of no consequence because the source must be finite and we are not interested in the solution at $r = 0$.

In this particular case we have assumed a source radiating into an unbounded space and so there will only be an outgoing wave. Furthermore it is sufficient to assume a simple harmonic fluctuation and the solution, using complex algebra, may be written

$$\mathbf{p} = \frac{\mathbf{A}}{r}e^{j(\omega t - kr)} \tag{3.11}$$

The particle velocity

$$\mathbf{u} = \frac{\mathbf{A}}{r} \cdot \frac{1}{\rho_0 c} \left(1 - j\frac{1}{kr}\right) e^{j(\omega t - kr)} \tag{3.12}$$

The pressure and particle velocity may be represented by the vectors in Fig. 3.2.

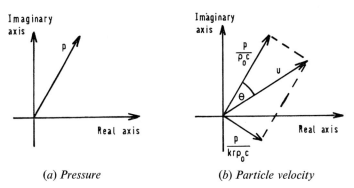

(a) Pressure (b) Particle velocity

Fig. 3.2. Vector diagrams for spherical radiation.

Alternatively the ratio of particle velocity to pressure may be written

$$\frac{\mathbf{u}}{\mathbf{p}} = \frac{1}{\rho_0 c} \left(1 - \frac{j}{kr}\right)$$

The real and imaginary parts of this expression are shown as a function of kr in Fig. 3.3.

The phase difference between p and u is a function of r varying from $\pi/2$ when r is very small to zero when r is very large.

In the general case

$$\theta = \tan^{-1}(1/kr) \tag{3.13}$$

and

$$\cos \theta = \frac{kr}{(1 + k^2 r^2)^{1/2}} \tag{3.14}$$

The particle velocity could be written

$$\mathbf{u} = \frac{\mathbf{A}}{\rho_0 c} \frac{(1 + k^2 r^2)^{1/2}}{kr^2} e^{j(\omega t - kr - \theta)} \tag{3.15}$$

We are now in a position to calculate the energy density and intensity associated with the outgoing wave, using equations (3.11) and (3.15) to provide the moduli of pressure and particle velocity for substitution into equations (2.57) and (2.58)

$$D(r) = \frac{1}{2}\left\{ \frac{|\mathbf{A}|^2}{2\rho_0 c^2}\frac{1+k^2 r^2}{k^2 r^4} + \frac{|\mathbf{A}|^2}{2\rho_0 c^2 r^2} \right\} = \frac{|\mathbf{A}|^2}{2\rho_0 c^2 r^2}\left(1 + \frac{1}{2k^2 r^2}\right) \quad (3.16)$$

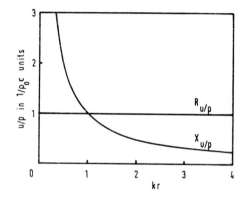

Fig. 3.3.   *The real and imaginary parts of the ratio of particle velocity to pressure radiated by a spherical source.*

The rms pressure is given by

$$\bar{p}^2 = \frac{|\mathbf{A}|^2}{2r^2} \quad (3.17)$$

therefore
$$D(r) = \frac{\bar{p}^2}{\rho_0 c^2}\left(1 + \frac{1}{2k^2 r^2}\right) \quad (3.18)$$

The intensity
$$I(r) = \frac{|\mathbf{A}|^2}{2\rho_0 c r^2} \quad (3.19)$$

$$= \frac{\bar{p}^2}{\rho_0 c} \quad (3.20)$$

The two components of the particle velocity, given in equation (3.12) merit a little more attention. The first is inversely proportional to r and is equal to the pressure divided by $\rho_0 c$. This is identical to the particle velocity in a one-dimensional progressive wave except for the term $1/r$. The second is inversely proportional to r^2 and is 90° out of phase with the pressure. This latter term may be thought of as gas attached to, and moving with, the source. It is significant very close to the source but negligible when $kr \gg 1$. It does not contribute in any way to the energy which is radiated away from the

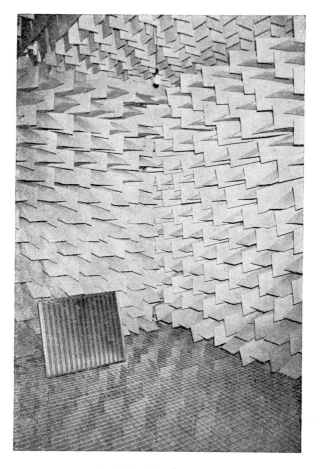

Fig. 3.4.   Anechoic room.

source, as we can see from equation (3.20), but close to the source it does contribute to the kinetic energy term in the expression for energy density.

Equation (3.19) tells us that the intensity is inversely proportional to r^2, which we deduced more simply from the physics of the situation in Chapter 1, and which is a statement of the inverse square law common in other branches of physics. The important point is that the intensity can be directly related to the pressure which is a measurable quantity.

In practice the relationship given in equation (3.20) holds for sources which are not spherically symmetric provided that measurements are not made close to the source. Intensity radiation patterns may be obtained from the pressure patterns and the total power output of the source may be calculated by integrating the intensity over the surface of an imaginary sphere of large radius. It is not possible to conduct such experiments literally in free space but the laboratory equivalent is an anechoic room. Ideally the working space should be as large as possible and there should be no hard reflecting surfaces. The walls, floor and ceiling are all covered with foam or mineral wool wedges designed to reflect the minimum possible amount of sound. Fig. 3.4 shows such a room with wedges 1 m long and a working section 5 by 3·5 by 3 m which provides reasonable free field conditions at frequencies above 100 Hz.

3.3   Radiation from a Simple Source

The simplest type of source for producing spherical waves is a small pulsating sphere. Although this might seem to have little practical significance it is useful in that many real radiators can be likened to a sphere if their dimensions are small compared with the wavelength of the sound which is being radiated.

The sphere can be defined in terms of its radius, a, and the radial velocity u_a at any point on the surface. Suppose that

$$u_a = U \cos \omega t$$

The complex form of this equation is

$$\mathbf{u}_a = U e^{j\omega t} \tag{3.21}$$

The fluid surrounding the sphere must at all times remain in contact with the surface. Therefore the particle velocity in the sound wave given by equation (3.12) with $r = a$ must be equal to the velocity u_a.

i.e.
$$U e^{j\omega t} = \frac{A}{a\rho_0 c}\left(1-j\frac{1}{ka}\right)e^{j(\omega t-ka)}$$

and
$$A = U\rho_0 c\,\frac{ka^2(ka+j)}{1+k^2a^2}(\cos ka+j\sin ka) \qquad (3.22)$$

A is now known and the radiated pressure can be found by substituting equation (3.22) into equation (3.11) and then taking the real part. In the particular case of the sphere being small compared with the radiated wavelengths we may write $ka \ll 1$ and then equation (3.22) reduces to
$$\mathbf{A} \simeq jU\rho_0 cka^2$$

Substitution into equation (3.11) gives
$$\mathbf{p} = \frac{jU\rho_0 cka^2}{r}e^{j(\omega t-kr)} \qquad (3.23)$$

The real part of this equation, which represents the actual pressure, is
$$p = -\frac{U\rho_0 cka^2}{r}\sin(\omega t-kr)$$

If the source strength is defined as the product of the velocity amplitude and the surface area

i.e.
$$Q = \int_s U dS$$
$$= 4\pi a^2 U \quad \text{for a sphere} \qquad (3.24)$$

then the expression for pressure can be re-written
$$\mathbf{p} = \frac{j\rho_0 ck}{4\pi r}Q\,e^{j(\omega t-kr)} \qquad (3.25)$$

which is extremely useful because, far from the source, it is found that the precise shape of the source is immaterial. So if the source strength can be defined, equation (3.25) can be used for obtaining the radiated pressure.

The effect of cutting the source and the infinite space in half by means of a rigid baffle is to leave the pressure field exactly as it was. It could, however, now be produced on one side of the baffle only by a pulsating hemisphere of strength $Q = 2\pi a^2 U$. Equation (3.25) must consequently be modified by a factor of 2 in order to give the expression relating pressure to source strength for a simple source located in an infinite baffle

i.e.
$$\mathbf{p} = \frac{j\rho_0 ck}{2\pi r} Q \, e^{j(\omega t - kr)} \qquad (3.26)$$

3.4 Radiation from a Piston

A flat piston located in an infinite baffle can be thought of as a set of simple sources. The pressure from one elemental source is

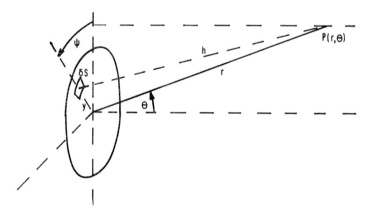

Fig. 3.5.   Radiation from a piston set in a rigid baffle.

given by equation (3.26) and the net effect is obtained by integrating over the whole surface area of the piston. If the piston is circular, it is best to use spherical coordinates as shown in Fig. 3.5.

Suppose that the piston is of radius a and that the whole surface is vibrating uniformly with a velocity $U \, e^{j\omega t}$. Then the elemental pressure at P produced by the elemental area δS is

$$\delta \mathbf{p} = \frac{j\rho_0 ck}{2\pi h} U e^{j(\omega t - kh)} \delta S \tag{3.27}$$

where $$\delta S = y \, \delta y \, \delta \psi$$

In order to carry out the integration it is necessary to express h in terms of the other variables. It may be shown geometrically that

$$h = (r^2 + y^2 - 2ry \sin \theta \cos \psi)^{1/2}$$

If P is far from the piston, then

$$h \simeq r - y \sin \theta \cos \psi$$

Upon substitution it is sufficient to write $h \simeq r$ in the amplitude term but the phase term is more critical and requires full substitution.

So $$\delta \mathbf{p} = \frac{j\rho_0 ck}{2\pi r} U e^{j(\omega t - kr + ky \sin \theta \cos \psi)} y \, \delta y \, \delta \psi$$

The total pressure is

$$\mathbf{p} = \frac{j\rho_0 ck}{2\pi r} U e^{j(\omega t - kr)} \int_0^a y \, dy \int_0^{2\pi} e^{jky \sin \theta \cos \psi} \, d\psi$$

The integration is carried out by expanding the exponential as a power series to give

$$\mathbf{p} = \frac{j\rho_0 ck}{2\pi r} U e^{j(\omega t - kr)} \int_0^a 2\pi y J_0(ky \sin \theta) \, dy$$

$$= \frac{j\rho_0 ck}{2\pi r} \pi a^2 U e^{j(\omega t - kr)} \left[\frac{2J_1(ka \sin \theta)}{ka \sin \theta} \right] \tag{3.28}$$

where $$J_0(x) = 1 - \frac{x^2}{2^2} + \frac{x^4}{2^2 4^2} \cdots$$

and $$J_1(x) = \frac{1}{2} \left(x - \frac{x^3}{2.4} + \frac{x^5}{2.4.4.6} \cdots \right)$$

J_0 and J_1 are known as the zero and first order Bessel functions and often cause some dismay due to lack of familiarity. In fact they are simply names for particular series in exactly the same way that sines and cosines are names for particular series. The functions have, of course, been calculated and it is only necessary to consult tables of Bessel functions to obtain values for J_0 and J_1.

The pressure produced by the piston of strength $\pi a^2 U$ is the same as for a hemisphere of equal strength except for the directivity factor

$$\left[\frac{2J_1(ka \sin \theta)}{ka \sin \theta}\right]$$

which is shown as a function of $ka \sin \theta$ in Fig. 3.6.

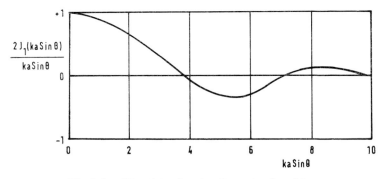

Fig. 3.6. *Directivity function for a circular piston.*

Along the axis through the piston the directivity function is 1 and the radiated pressure is directly proportional to k, *i.e.* proportional to frequency. At low frequencies then, the pressure is low but is almost uniform over the whole hemisphere because $ka \sin \theta$ is also small, even with $\theta = \pi/2$. At higher frequencies the pressure along the axis increases but pressure lobes occur at other values of θ. This is illustrated in Fig. 3.7.

The theory relating to radiation from a piston is of some relevance to the radiation from loudspeakers. A loudspeaker cone is not, of course, rigid and flexure upsets the situation to some extent but at least it can be concluded that high frequency speakers must be of small size if they are not to be too directional. Conversely low frequency speakers can be larger and still remain omnidirectional and indeed must be larger if they are to create a sufficiently high pressure.

3.5   Reaction on a Piston

The reaction of the air back on the piston is important because it is necessary to include this term when trying to calculate the velocity of the piston produced by a known input force. The term is additive to the normal reaction of the piston, possibly including mass or stiffness terms, which exists even when the piston is operating in a vacuum. When the piston moves, the total pressure in the vicinity of any elemental area is due to direct reaction on that element plus the radiated pressure from every other elemental area.

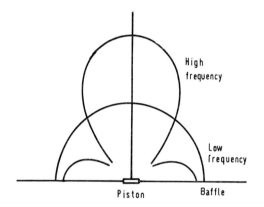

Fig. 3.7.   Radiation patterns for a circular piston mounted in an infinite baffle.

Fig. 3.8 shows the pressure produced at the elemental area $\delta S'$ due to the movement of the elemental area δS.

The total acoustic pressure p at $\delta S'$ is again given by

$$\mathbf{p} = \iint \frac{j\rho_0 ck}{2\pi r} \, U \, e^{j(\omega t - kr)} \, dS$$

And the total force acting on the piston

$$\mathbf{F} = \iint \mathbf{p} \, dS'$$

$$= \frac{j\rho_0 ck}{2\pi} U e^{j\omega t} \iint dS' \iint \frac{e^{-jkr}}{r} \, dS$$

This integral can be simplified if it is noted that each pair of elements interact. Therefore we should integrate only once between any two elements and multiply the result by 2. This can be arranged by specifying that the element δS should always be within the circle of radius y.

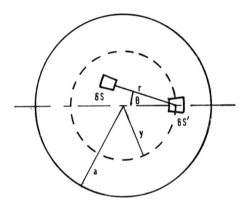

Fig. 3.8.   *Pressure at the surface of the piston.*

Then
$$\delta S = r\,\delta\theta\,\delta r$$

$$\delta S' = y\,\delta y\,\delta\psi$$

$$\mathbf{F} = \frac{j\rho_0 ck}{\pi} U\,e^{j\omega t} \int_0^a y\,dy \int_0^{2\pi} d\psi \int_{-\pi/2}^{+\pi/2} d\theta \int_0^{2y\cos\theta} e^{-jkr}\,dr$$

$$= \rho_0 c\pi a^2 U\,e^{j\omega t}[R_1(2ka)+jX_1(2ka)]$$

where
$$R_1(x) = \frac{x^2}{2.4} - \frac{x^4}{2.4^2.6} + \frac{x^6}{2.4^2.6^2.8} \cdots$$

and
$$X_1(x) = \frac{4}{\pi}\left(\frac{x}{3} - \frac{x^3}{3^2.5} + \frac{x^5}{3^2.5^2.7}\cdots\right)$$

The force acting on the piston must be equal to the force exerted by the piston and the radiation impedance, which is defined as the ratio of the piston force to the piston velocity, is given by

$$\mathbf{Z}_R = \frac{\mathbf{F}}{U\,e^{j\omega t}} = \rho_0 c\pi a^2[R_1(2ka)+jX_1(2ka)]$$

In the particular case where the piston is small compared with the wavelength of sound which is being radiated, *i.e.* $ka \ll 1$

then
$$R_1(2ka) \simeq \frac{k^2 a^2}{2}; \quad X_1(2ka) \simeq \frac{8ka}{3\pi}$$

$$\mathbf{Z}_R = \rho_0 c \pi a^2 \left[\frac{k^2 a^2}{2} + j \frac{8ka}{3\pi} \right] \tag{3.29}$$

The real part of $\mathbf{Z}_R$ is due to the reaction of the true radiated pressure, but the imaginary part is effectively a mass loading of the piston by the air in its immediate vicinity.

We shall be using the expression for $\mathbf{Z}_R$ given by equation (3.29) in Chapter 5 when discussing the loading at the end of a pipe which is radiating into free space and at the neck of a Helmholtz resonator. Indeed part of it has already been used indirectly in Chapter 2 when it was stated that an end correction of $8a/3\pi$ is normally added to the length of a pipe which has an open flanged end. In these instances the piston is not a solid but is the thin vibrating layer of gas right at the mouth of the pipe.

3.6   Sound Waves in a Rectangular Enclosure

The behaviour of sound in a rectangular enclosure is important because it is relevant to room acoustics. Rooms are not, of course, necessarily rectangular but the wave equation can be solved exactly in this particular case and the results can then be generalised to include rooms of other shapes.

It is assumed that all of the walls are smooth, rigid and perfectly reflecting and, as shown in Fig. 3.9, one corner is chosen as the origin. The dimensions of the enclosure are l_x, l_y and l_z.

The wave equation in Cartesian coordinates is

$$\frac{\partial^2 p}{\partial x^2} + \frac{\partial^2 p}{\partial y^2} + \frac{\partial^2 p}{\partial z^2} = \frac{1}{c^2} \frac{\partial^2 p}{\partial t^2} \tag{3.30}$$

The general harmonic solution for a plane wave in three dimensions may be written

$$\mathbf{p} = \frac{\mathbf{A}}{8} (e^{-jk_x x} + \mathbf{B}\, e^{jk_x x})(e^{-jk_y y} + \mathbf{C}\, e^{jk_y y})(e^{-jk_z z} + \mathbf{D}\, e^{jk_z z})\, e^{j\omega t} \tag{3.31}$$

Substitution of this solution into equation (3.30) produces the requirement that

$$k = \frac{\omega}{c} = (k_x^2 + k_y^2 + k_z^2)^{1/2} \tag{3.32}$$

This is not very remarkable, merely defining the relationship between the plane wave and its three components. k_x/k, k_y/k and k_z/k are sometimes called the direction cosines of the wave on the three axes

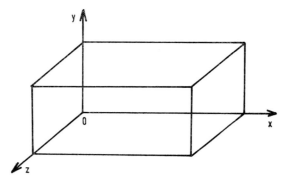

Fig. 3.9.   Co-ordinates of rectangular enclosure.

x, y and z. The only point to note is that the corresponding wavelengths in the three directions are, in general, greater than the wavelength of the plane wave.

The particle velocity also has three components where, for example,

$$\mathbf{u}_x = -\frac{1}{\rho_0} \int \frac{\partial \mathbf{p}}{\partial x} \partial t$$

The rigid boundaries impose the condition that the particle velocity normal to each boundary should be zero at all times.

i.e. $\mathbf{u}_x = 0$ at $x = 0$ and $x = l_x$

$\mathbf{u}_y = 0$ at $y = 0$ and $y = l_y$

$\mathbf{u}_z = 0$ at $z = 0$ and $z = l_z$

Imposing the set of conditions at x, y, $z = 0$ simplifies equation (3.31) considerably by reducing it to the form

$$\mathbf{p} = \mathbf{A} \cos k_x x \cos k_y y \cos k_z z \, e^{j\omega t} \qquad (3.33)$$

Imposing the other set of conditions gives the following requirements

$$k_x = \frac{n_x \pi}{l_x}; \qquad n_x = 0, 1, 2, 3 \ldots$$

$$k_y = \frac{n_y \pi}{l_y}; \qquad n_y = 0, 1, 2, 3 \ldots$$

$$k_z = \frac{n_z \pi}{l_z}; \qquad n_z = 0, 1, 2, 3 \ldots \qquad (3.34)$$

Finally, substitution of equation (3.34) into equation (3.32) yields the important relationship.

$$\frac{\omega}{c} = \pi \left[\left(\frac{n_x}{l_x}\right)^2 + \left(\frac{n_y}{l_y}\right)^2 + \left(\frac{n_z}{l_z}\right)^2 \right]^{1/2}$$

therefore
$$f = \frac{c}{2} \left[\left(\frac{n_x}{l_x}\right)^2 + \left(\frac{n_y}{l_y}\right)^2 + \left(\frac{n_z}{l_z}\right)^2 \right]^{1/2} \qquad (3.35)$$

The rectangular room will sustain resonant or stationary waves at the frequencies dictated by equation (3.35). For each resonant frequency there is a particular wave pattern which is obtained by substituting the appropriate values for k_x, k_y and k_z into equation (3.33). This tells us that a pressure antinode occurs in each corner of the room for every standing wave pattern. Because of this it is generally recommended that a loudspeaker should be placed in the corner of a room for then it will excite as many room resonances as can exist. Elsewhere it would tend to be selective, exciting only those waves which had pressure antinodes in its vicinity. In order to give some idea of the significance of these standing waves the first 25 natural frequencies of a rectangular room of dimensions 6 by 5 by 3 m have been calculated and are shown in Table 3.1. The speed of sound in air is assumed to be 343 m.sec^{-1}.

The first few resonances are well separated from each other in frequency but it is obvious that the number of resonances in any given frequency band increases as the frequency increases. This is

TABLE 3.1

THE FIRST 25 NATURAL FREQUENCIES OF A RECTANGULAR
ROOM IN WHICH $l_x = 6$ m; $l_y = 5$ m; $l_z = 3$ m

n_x	n_y	n_z	f in Hz
1	0	0	29
0	1	0	34
1	1	0	45
0	0	1	57
2	0	0	57
1	0	1	64
0	1	1	67
2	1	0	67
0	2	0	69
1	1	1	72
1	2	0	74
2	0	1	80
3	0	0	86
2	1	1	88
0	2	1	89
2	2	0	89
3	1	0	92
1	2	1	92
3	0	1	103
0	3	0	103
2	2	1	106
3	1	1	108
3	2	0	109
0	0	2	114
4	0	0	114

defined as modal density and we may obtain a better idea of the way in which it varies by rewriting equation (3.35) in the form

$$f^2 = \left(\frac{cn_x}{2l_x}\right)^2 + \left(\frac{cn_y}{2l_y}\right)^2 + \left(\frac{cn_z}{2l_z}\right)^2 \qquad (3.36)$$

We know that there is a value of f for every combination of the independently variable integers n_x, n_y and n_z. Equation (3.36) can therefore be represented graphically by the rectangular lattice shown in Fig. 3.10.

Each intersection, shown by a dot in Fig. 3.10, represents one solution to equation (3.36) and the appropriate frequency is equal to the length of the vector drawn from the origin to the intersection. The number of resonances N below a certain frequency f is equal to the number of intersections, or dots, contained within the volume bounded by the three coordinate planes and a spherical surface of radius f centred on the origin.

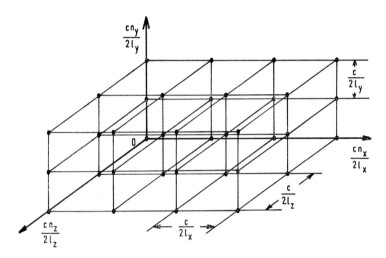

Fig. 3.10. *Graphical representation of equation* (3.36).

Now that volume

$$= \tfrac{4}{3}\,\pi f^3/8$$

$$= \frac{\pi f^3}{6}$$

and the number of intersections per unit volume of the lattice

$$= \frac{2l_x}{c}\cdot\frac{2l_y}{c}\cdot\frac{2l_z}{c}$$

$$= \frac{8V}{c^3}$$

where V is the volume of the room.

Therefore the number of resonances occurring below the frequency f is given by

$$N = \frac{\pi f^3}{6} \times \frac{8V}{c^3}$$

$$= \frac{4\pi V f^3}{3c^3} \tag{3.37}$$

TABLE 3.2

NUMBER OF MODES AND MODAL DENSITY FOR A
RECTANGULAR ROOM OF SIZE 6 BY 5 BY 3 m

Frequency in Hz	Modal density	Number of modes below this frequency
63	0·1	2
125	0·4	18
250	1·7	146
500	7·0	1170
1000	28	9340
2000	112	74,600
4000	448	597,000
8000	1790	4,770,000

The modal density can be obtained by differentiating equation (3.37) which gives

$$\delta N = \frac{4\pi V f^2}{c^3}\, \delta f \tag{3.38}$$

The modal density increases according to the square of the frequency which bears out the general conclusion drawn from the example. In fact, equations (3.37) and (3.38) are not very accurate at low frequencies, i.e. when N is small anyway, because the points lying in the f_x, f_y and f_z planes in Fig. 3.9 are ignored, but at high frequencies they are very good.

Values of N and δN at a number of different frequencies are shown in Table 3.2 for the rectangular room used in the previous example. For modal density, the bandwidth δf is taken to be 1 Hz.

3.7 Room Acoustics

Normally in room acoustics we are interested in the response of the room at all frequencies up to at least 8000 Hz. The example in the previous section shows that, although standing waves are set up, there are going to be so many in the average room in this frequency range, that the individual consideration of each mode becomes an impossible task. An extra complication is that the wave equation can only be solved exactly for rectangular, cylindrical and spherical rooms and few rooms are so simple.

It is better to acknowledge the existence of standing waves but to assume that the modal density will be high, except at very low frequencies in small rooms. The modal density can always be checked by using equation (3.38). There is now no restriction on the shape of the room because only the volume is involved. As an example, the modal density in a small concert hall of volume 30,000 m³ at a frequency of 100 Hz is approximately 9 modes Hz⁻¹. It then follows that a large number of modes will always be excited and the different standing wave patterns will tend to average out, so giving a reasonably uniform rms pressure throughout the enclosure. The exception is close to the boundaries where all of the standing wave patterns have pressure antinodes and the average rms pressure close to the walls is double the rms pressure elsewhere in the enclosure.

In order to find the energy density of any standing wave pattern in the enclosure we may use equation (3.33) to give the modulus of the pressure and the moduli of the three components of particle velocity $\mathbf{u}_x$, $\mathbf{u}_y$ and $\mathbf{u}_z$.

i.e. if
$$\mathbf{u}_x = \frac{k_x}{j\rho_0 ck} A \sin k_x x \cos k_y y \cos k_z z \, e^{j\omega t}$$

then
$$|\mathbf{p}| = |A| \cos k_x x \cos k_y y \cos k_z z$$

and
$$|\mathbf{u}_x| = \frac{k_x}{\rho_0 ck} |A| \sin k_x x \cos k_y y \cos k_z z$$

Noting that
$$|\mathbf{u}|^2 = |\mathbf{u}_x|^2 + |\mathbf{u}_y|^2 + |\mathbf{u}_z|^2$$

we can obtain by substitution into equation (2.57)

$$D(x,y,z) = \frac{|\mathbf{A}|^2}{4\rho_0 c^2} \left(\frac{k_x^2}{k^2} \cos^2 k_y y \cos^2 k_z z + \frac{k_y^2}{k^2} \cos^2 k_x x \cos^2 k_z z \right.$$
$$\left. + \frac{k_z^2}{k^2} \cos^2 k_x x \cos^2 k_y y \right)$$

which, upon averaging in space reduces to

$$D = \frac{|\mathbf{A}|^2}{16\rho_0 c^2} \tag{3.39}$$

The mean square pressure

$$\bar{p}^2(x,y,z) = \frac{|\mathbf{A}|^2}{2} \cos^2 k_x x \cos^2 k_y y \cos^2 k_z z$$

and, averaging in space,

$$\bar{p}^2 = \frac{|\mathbf{A}|^2}{16} \tag{3.40}$$

therefore

$$D = \frac{\bar{p}^2}{\rho_0 c^2} \tag{3.41}$$

which is the same relationship as was obtained for a one-dimensional standing wave, except that it has been necessary to average the rms pressure in three-dimensional space.

If sufficient modes are always excited, the space averaging process is in effect carried out by the room. The rms pressure can theoretically be measured at any point, except close to the boundaries, and is related to the energy density in the room by equation (3.41). In practice it is better to average the rms pressure from measurements made at several different points because the pressure is constant throughout the enclosure only when the modal density is infinite, but there is not normally much variation. A sound field in which there is very little variation in pressure, and therefore energy density, and in which sound energy is travelling equally in all directions, is said to be diffuse.

By definition the total intensity in a standing wave is zero but this is only because there are equal and opposite components which cancel each other out. The surface of the enclosure is subjected to an incident sound intensity and it is necessary to relate this intensity to the energy density within the room so that we can take into account the possibility of absorption of sound at the boundaries.

In Fig. 3.11 the shaded element is three-dimensional and has a volume given by

$$\delta V = r^2 \sin \theta \; \delta\theta \; \delta r \; \delta\varphi$$

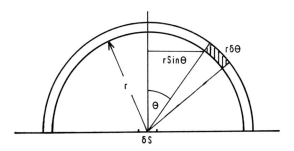

Fig. 3.11.   *Co-ordinate system for energy striking a boundary.*

The energy density is assumed to be constant throughout the room and so the energy contained within this element is $D \; \delta V$. This energy must be travelling equally in all directions and so the energy leaving the element per unit solid angle $= D \; \delta V/4\pi$. We are interested in the energy striking the elemental area δS, and the solid angle subtended by $\delta S = \delta S \cos \theta/r^2$.

The energy originating from δV which will strike δS

$$= \frac{\delta S \cos \theta}{r^2} \cdot \frac{D \; \delta V}{4\pi}$$

$$= \frac{D \; \delta S}{4\pi} \sin \theta \cos \theta \; \delta\theta \; \delta r \; \delta\varphi$$

To find the energy which strikes δS in unit time as a function of the angle of incidence θ, we must integrate cylindrically, *i.e.* integrate φ from 0 to 2π, and radially for a distance numerically equal to the velocity of sound, *i.e.* integrate r from 0 to c. Then dividing through by δS gives the intensity at any angle θ, *i.e.*

$$I_\theta = \frac{D}{4\pi} \sin \theta \cos \theta \, \delta\theta \int_0^c dr \int_0^{2\pi} d\varphi$$

$$= \frac{Dc}{2} \sin \theta \cos \theta \, \delta\theta \tag{3.42}$$

Finally to obtain the total intensity we can integrate θ from 0 to $\pi/2$ *i.e.*

$$I = \frac{Dc}{2} \int_0^{\pi/2} \sin \theta \cos \theta \, d\theta$$

$$= \frac{Dc}{4} \tag{3.43}$$

Equation (3.43) gives the total intensity from all angles of incidence striking any surface of an enclosure containing a diffuse sound field. Equation (3.41) gives the energy density in terms of the average mean square pressure and we may summarise by writing

$$I = \frac{Dc}{4} = \frac{\bar{p}^2}{4\rho_0 c} \tag{3.44}$$

CHAPTER 4

Transmission Through Layered Media

4.1 Characteristic Impedance and Specific Acoustic Impedance

In acoustics the concept of impedance is very useful and a number of different forms of impedance have been defined and are in use. This can lead to some confusion but since different forms are appropriate to different topics it is essential that we differentiate between them and understand their precise meaning.

In this chapter we are concerned with one-dimensional waves striking flat surfaces, which are assumed to be of infinite extent. Surface area is therefore of no importance and the most appropriate form of impedance is the specific acoustic impedance, which is, by definition,

$$z = \frac{\text{pressure}}{\text{particle velocity}} = \frac{p}{u} \qquad (4.1)$$

In general two waves will exist in a medium, one travelling in each direction, and so the pressure and particle velocity in equation (4.1) will include terms from both waves and will be a function of position. The particular function is determined by boundary conditions, as, for example, at a rigid wall where the particle velocity normal to the wall must clearly be zero. This is the same as saying that the specific acoustic impedance of that component of the wave normal to the wall is infinite.

In the particular case of a single one-dimensional wave travelling through a medium of infinite extent the pressure and particle velocity in equation (4.1) only contain single terms and we know from equation (2.34) that the relationship is a simple constant

i.e.
$$z_0 = \frac{\mathbf{p}}{\mathbf{u}} = \pm \rho_0 c \qquad (4.2)$$

The plus sign refers to a wave travelling in the positive-x direction while the minus sign refers to a wave travelling in the negative-x direction and is merely a matter of definition. In either case the constant is purely dependent on the properties of the medium and z_0 is known as the characteristic impedance. The velocity, c, and therefore the wave number, k, are in general complex, with the imaginary part accounting for absorption as the wave travels through the medium. In most cases, however, the absorption is small and, if neglected, the velocity and the wave number become wholly real. If this is so then the characteristic impedance is also wholly real.

4.2   Normal Incidence Transmission Through Two Layers

Suppose that a plane sound wave is normally incident upon a boundary separating two media. Upon striking the boundary some

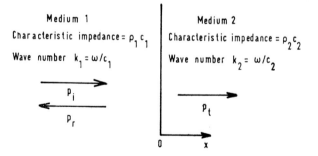

Fig. 4.1.   *Waves on either side of a boundary separating two media.*

of the incident wave will be reflected and some will be transmitted. These waves may be represented in their exponential form as follows

$$\mathbf{p}_i = A_1 \, e^{j(\omega t - k_1 x)} \qquad (4.3)$$

$$\mathbf{p}_r = \mathbf{B}_1 \, e^{j(\omega t + k_1 x)} \tag{4.4}$$

$$\mathbf{p}_t = \mathbf{A}_2 \, e^{j(\omega t - k_2 x)} \tag{4.5}$$

$\mathbf{A}_1$, $\mathbf{B}_1$ and $\mathbf{A}_2$ are complex in order to allow for phase differences between the three waves.

If we consider each of the three waves separately we can divide the pressure by the appropriate characteristic impedance to obtain the particle velocity

i.e.
$$\mathbf{u}_i = \frac{\mathbf{p}_i}{\rho_1 c_1} \tag{4.6}$$

$$\mathbf{u}_r = -\frac{\mathbf{p}_r}{\rho_1 c_1} \tag{4.7}$$

$$\mathbf{u}_t = \frac{\mathbf{p}_t}{\rho_2 c_2} \tag{4.8}$$

At the boundary, two conditions must be satisfied.

1. *Continuity of pressure*

$$(\mathbf{p}_i + \mathbf{p}_r)_{x=0} = (\mathbf{p}_t)_{x=0}$$

i.e.
$$\mathbf{A}_1 + \mathbf{B}_1 = \mathbf{A}_2 \tag{4.9}$$

2. *Continuity of particle velocity*

$$(\mathbf{u}_i + \mathbf{u}_r)_{x=0} = (\mathbf{u}_t)_{x=0}$$

i.e.
$$\frac{\mathbf{A}_1 - \mathbf{B}_1}{\rho_1 c_1} = \frac{\mathbf{A}_2}{\rho_2 c_2} \tag{4.10}$$

$\mathbf{A}_2$ may be eliminated from equations (4.9) and (4.10) to give the reflected wave in terms of the incident wave.

i.e.
$$\frac{\mathbf{B}_1}{\mathbf{A}_1} = \frac{\rho_2 c_2 - \rho_1 c_1}{\rho_2 c_2 + \rho_1 c_1} \tag{4.11}$$

or $\mathbf{B}_1$ may be eliminated to give the transmitted wave

i.e.
$$\frac{\mathbf{A}_2}{\mathbf{A}_1} = \frac{2 \rho_2 c_2}{\rho_2 c_2 + \rho_1 c_1} \tag{4.12}$$

As an alternative both of the boundary conditions could have been combined into the single condition that the specific acoustic impedance is continuous at the boundary, and is equal to z_s.

Then
$$\left(\frac{\mathbf{p}_i+\mathbf{p}_r}{\mathbf{u}_i+\mathbf{u}_r}\right)_{x=0} = z_s = \left(\frac{\mathbf{p}_t}{\mathbf{u}_t}\right)_{x=0}$$

i.e.
$$\frac{\mathbf{A}_1+\mathbf{B}_1}{\mathbf{A}_1-\mathbf{B}_1}\rho_1 c_1 = z_s = \rho_2 c_2 \tag{4.13}$$

This tells us nothing about the transmitted wave since $\mathbf{A}_2$ has been eliminated, but it gives a useful solution in the more general case of a wave being reflected from a surface of impedance z_s. Equation (4.13) may be rearranged into the form

$$\frac{\mathbf{B}_1}{\mathbf{A}_1} = \frac{z_s-\rho_1 c_1}{z_s+\rho_1 c_1} \tag{4.14}$$

Returning to equations (4.11) and (4.12), we may calculate the sound intensity reflection and transmission coefficients as follows

$$\alpha_r = \frac{I_r}{I_i} = \frac{|\mathbf{B}_1|^2/2\rho_1 c_1}{|\mathbf{A}_1|^2/2\rho_1 c_1} = \left|\frac{\mathbf{B}_1}{\mathbf{A}_1}\right|^2$$
$$= \left(\frac{\rho_2 c_2-\rho_1 c_1}{\rho_2 c_2+\rho_1 c_1}\right)^2 \tag{4.15}$$

$$\alpha_t = \frac{I_t}{I_i} = \frac{|\mathbf{A}_2|^2/2\rho_2 c_2}{|\mathbf{A}_1|^2/2\rho_1 c_1} = \frac{\rho_1 c_1}{\rho_2 c_2}\left|\frac{\mathbf{A}_2}{\mathbf{A}_1}\right|^2$$
$$= \frac{4\rho_1 c_1\rho_2 c_2}{(\rho_2 c_2+\rho_1 c_1)^2} \tag{4.16}$$

As an example, consider the transmission of a sound wave from air to water. Then $\rho_1 c_1 = 415$ kg m^{-2} sec^{-1} and $\rho_2 c_2 = 1,480,000$ kg m^{-2} sec^{-1}. $\rho_2 c_2$ is so much greater than $\rho_1 c_1$ that equation (4.16) may be written

$$\alpha_t \simeq \frac{4\rho_1 c_1}{\rho_2 c_2} = 1.12\times 10^{-3}$$

So only 0·112% of the incident energy is transmitted into the water.

In terms of pressures, we find from equation (4.11) that, at the surface, the reflected pressure is in phase with and very nearly equal to the incident pressure and so there is a pressure doubling effect. The transmitted pressure is almost twice the incident pressure but the particle velocity in the transmitted wave is very much smaller than that in the incident wave which accounts for the small transmission of intensity. The pressure doubling at the surface occurs whenever a wave tries to pass from a medium of low characteristic impedance to another of higher impedance, *e.g.* when a sound wave in a room strikes a solid wall.

If, in our example, we had considered a wave passing from water into air, then the transmission coefficient would have been exactly the same. The reflected pressure would have been very nearly equal in magnitude to the incident pressure but exactly out of phase and so there would be a pressure cancellation. The amplitude of the transmitted pressure would then be very small compared with the incident pressure. This situation arises whenever a wave tries to pass from a medium of high characteristic impedance to another of low characteristic impedance.

4.3   Oblique Incidence Transmission Through Two Fluid Layers

The preceding section did not specifically apply to fluid layers because at normal incidence only longitudinal waves can be induced. Oblique incidence waves, however, will also tend to induce shear waves if the medium can support them. The situation therefore becomes very much more complicated in the case of a solid and in the case of a very viscous fluid. In this section we shall restrict ourselves to non-viscous fluids, so that only longitudinal waves need be considered.

Since we are dealing with wavefronts, Snell's laws of reflection and refraction must be obeyed so that

$$\frac{\sin \theta_2}{\sin \theta_1} = \frac{c_2}{c_1} = \frac{k_1}{k_2} \qquad (4.17)$$

The three waves may still be represented exponentially but the exponent must now also contain a term in y

i.e.
$$\mathbf{p}_i = \mathbf{A}_1\, e^{j(\omega t - k_1 x \cos\theta_1 - k_1 y \sin\theta_1)} \tag{4.18}$$

$$\mathbf{p}_r = \mathbf{B}_1\, e^{j(\omega t + k_1 x \cos\theta_1 - k_1 y \sin\theta_1)} \tag{4.19}$$

$$\mathbf{p}_t = \mathbf{A}_2\, e^{j(\omega t - k_2 x \cos\theta_2 - k_2 y \sin\theta_2)} \tag{4.20}$$

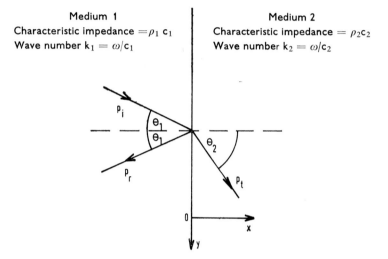

Medium 1	Medium 2
Characteristic impedance $=\rho_1 c_1$	Characteristic impedance $= \rho_2 c_2$
Wave number $k_1 = \omega/c_1$	Wave number $k_2 = \omega/c_2$

Fig. 4.2.   Oblique incidence waves on either side of a boundary separating two media.

In fact since $k_2 \sin\theta_2 = k_1 \sin\theta_1$, the y term is the same in all three waves and always cancels out. The particle velocity associated with each wave is obtained as before by dividing the pressure by the appropriate characteristic impedance.

At the boundary two conditions must be satisfied.

1. *Continuity of pressure*

$$(\mathbf{p}_i + \mathbf{p}_r)_{x=0} = (\mathbf{p}_t)_{x=0}$$

i.e.
$$\mathbf{A}_1 + \mathbf{B}_1 = \mathbf{A}_2 \tag{4.21}$$

2. *Continuity of particle velocity normal to the surface*

$$(\mathbf{u}_i \cos\theta_1 + \mathbf{u}_r \cos\theta_1)_{x=0} = (\mathbf{u}_t \cos\theta_2)_{x=0}$$

i.e.
$$\frac{(A_1 - B_1) \cos \theta_1}{\rho_1 c_1} = \frac{A_2 \cos \theta_2}{\rho_2 c_2} \tag{4.22}$$

Continuity of particle velocity tangential to the surface is not required since we have assumed that there is no viscosity.

Either A_2 or B_1 may be eliminated from equations (4.21) and (4.22) to give the reflected and transmitted waves *i.e.*

$$\frac{B_1}{A_1} = \frac{(\rho_2 c_2 \cos \theta_1 / \cos \theta_2) - \rho_1 c_1}{(\rho_2 c_2 \cos \theta_1 / \cos \theta_2) + \rho_1 c_1} \tag{4.23}$$

and
$$\frac{A_2}{A_1} = \frac{(2\rho_2 c_2 \cos \theta_1 / \cos \theta_2)}{(\rho_2 c_2 \cos \theta_1 / \cos \theta_2 + \rho_1 c_1)} \tag{4.24}$$

Alternatively we could specify continuity of specific acoustic impedance normal to the surface, and if this is designated z_s then, in the general case

$$\frac{B_1}{A_1} = \frac{z_s \cos \theta_1 - \rho_1 c_1}{z_s \cos \theta_1 + \rho_1 c_1} \tag{4.25}$$

The intensity reflection and transmission coefficients have now become a function of angle of incidence. For $c_2 > c_1$ the limiting case occurs when $\sin \theta_2 = 1$ which indicates total reflection and so there is no transmission at that angle or at greater angles of incidence. Returning to our earlier example of waves passing from air to water total reflection occurs with angles of incidence greater than about $13\cdot4°$ to the normal. For sound passing from water to air, however, all waves emerge at an angle of less than $13\cdot4°$ to the normal. This example cannot be extended to solid boundaries because shear waves have been neglected.

4.4 Normal Incidence Transmission Through Three Layers

Any single wavefront striking the first boundary would produce a reflected and transmitted wave. The latter would then produce multiple reflections and transmissions at the two boundaries. Fortunately it is sufficient to suppose that there is only one wave in each direction in medium 1 and medium 2 and one wave in medium 3 since, provided that the boundary conditions are satisfied, these will include all of the individual components. The five waves will have the following pressures and particle velocities.

$$\mathbf{p}_{i1} = \mathbf{A}_1 \, e^{j(\omega t - k_1 x)} \qquad \mathbf{u}_{i1} = \frac{\mathbf{p}_{i1}}{\rho_1 c_1} \qquad (4.26)$$

$$\mathbf{p}_{r1} = \mathbf{B}_1 \, e^{j(\omega t + k_1 x)} \qquad \mathbf{u}_{r1} = -\frac{\mathbf{p}_{r1}}{\rho_1 c_1} \qquad (4.27)$$

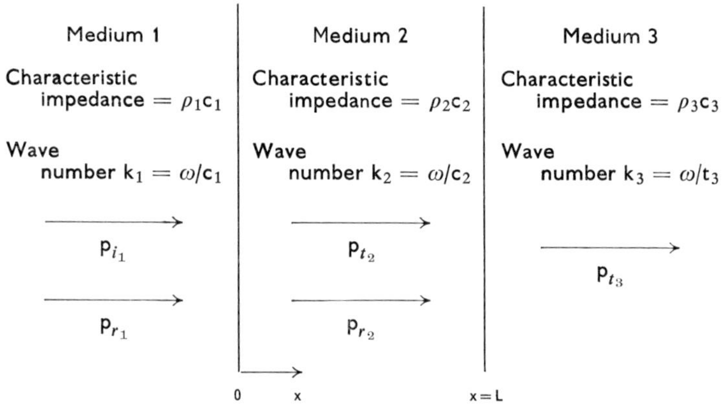

Medium 1	Medium 2	Medium 3
Characteristic impedance $= \rho_1 c_1$	Characteristic impedance $= \rho_2 c_2$	Characteristic impedance $= \rho_3 c_3$
Wave number $k_1 = \omega/c_1$	Wave number $k_2 = \omega/c_2$	Wave number $k_3 = \omega/t_3$
$\longrightarrow$ p_{i_1}	$\longrightarrow$ p_{t_2}	$\longrightarrow$ p_{t_3}
$\longrightarrow$ p_{r_1}	$\longrightarrow$ p_{r_2}	

$$\begin{array}{ccc} & \xrightarrow{\quad} & \\ 0 & x & x = L \end{array}$$

Fig. 4.3. Waves in three layers.

$$\mathbf{p}_{t2} = \mathbf{A}_2 \, e^{j(\omega t - k_2 x)} \qquad \mathbf{u}_{t2} = \frac{\mathbf{p}_{t2}}{\rho_2 c_2} \qquad (4.28)$$

$$\mathbf{p}_{r2} = \mathbf{B}_2 \, e^{j(\omega t + k_2 x)} \qquad \mathbf{u}_{r2} = -\frac{\mathbf{p}_{r2}}{\rho_2 c_2} \qquad (4.29)$$

$$\mathbf{p}_{t3} = \mathbf{A}_3 \, e^{j(\omega t - k_3(x-L))} \qquad \mathbf{u}_{t3} = \frac{\mathbf{p}_{t3}}{\rho_3 c_3} \qquad (4.30)$$

The conditions on pressure and particle velocity must now be satisfied at each boundary.
At $x = 0$

(1) *Continuity of pressure*

$$(\mathbf{p}_{i1} + \mathbf{p}_{r1})_{x=0} = (\mathbf{p}_{t2} + \mathbf{p}_{r2})_{x=0}$$

i.e. $$\mathbf{A}_1 + \mathbf{B}_1 = \mathbf{A}_2 + \mathbf{B}_2 \qquad (4.31)$$

(2) *Continuity of particle velocity*

$$(\mathbf{u}_{i1}+\mathbf{u}_{r1})_{x=0} = (\mathbf{u}_{t2}+\mathbf{u}_{r2})_{x=0}$$

i.e.

$$\frac{\mathbf{A}_1-\mathbf{B}_1}{\rho_1 c_1} = \frac{\mathbf{A}_2-\mathbf{B}_2}{\rho_2 c_2} \qquad (4.32)$$

At $x = L$

(3) *Continuity of pressure*

$$(\mathbf{p}_{t2}+\mathbf{p}_{r2})_{x=L} = (\mathbf{p}_{t3})_{x=L}$$

i.e.

$$\mathbf{A}_2\,e^{-jk_2 L}+\mathbf{B}_2\,e^{jk_2 L} = \mathbf{A}_3 \qquad (4.33)$$

(4) *Continuity of particle velocity*

$$(\mathbf{u}_{t2}+\mathbf{u}_{r2})_{x=L} = (\mathbf{u}_{t3})_{x=L}$$

i.e.

$$\frac{\mathbf{A}_2\,e^{-jk_2 L}-\mathbf{B}_2\,e^{jk_2 L}}{\rho_2 c_2} = \frac{\mathbf{A}_3}{\rho_3 c_3} \qquad (4.34)$$

Clearly equations (4.31) to (4.34) can be solved to give any of the reflected or transmitted waves in terms of the incident wave but in a three layer system it is most probable that we shall require the wave transmitted into medium 3. We shall, therefore, only attempt to find $\mathbf{A}_3/\mathbf{A}_1$. $\mathbf{B}_1$ may be eliminated from equations (4.31) and (4.32) to give

$$\mathbf{A}_1 = \frac{\mathbf{A}_2+\mathbf{B}_2}{2}+\frac{\rho_1 c_1}{\rho_2 c_2}\cdot\frac{\mathbf{A}_2-\mathbf{B}_1}{2} \qquad (4.35)$$

Both $\mathbf{A}_2$ and $\mathbf{B}_2$ may be found separately by combining equations (4.33) and (4.34).

$$\mathbf{A}_2 = \frac{\mathbf{A}_3}{2}\left(1+\frac{\rho_2 c_2}{\rho_3 c_3}\right)e^{jk_2 L} \qquad (4.36)$$

$$\mathbf{B}_2 = \frac{\mathbf{A}_3}{2}\left(1-\frac{\rho_2 c_2}{\rho_3 c_3}\right)e^{-jk_2 L} \qquad (4.37)$$

Substituting for $\mathbf{A}_2$ and $\mathbf{B}_2$ in equation (4.35) and rearranging gives

$$\frac{\mathbf{A}_1}{\mathbf{A}_3} = \frac{1}{2}\left(1+\frac{\rho_1 c_1}{\rho_3 c_3}\right)\cos k_2 L+j\frac{1}{2}\left(\frac{\rho_2 c_2}{\rho_3 c_3}+\frac{\rho_1 c_1}{\rho_2 c_2}\right)\sin k_2 L \qquad (4.38)$$

The intensity transmission coefficient is given by

$$\alpha_t = \frac{|A_3|^2/2\rho_3 c_3}{|A_1|^2/2\rho_1 c_1}$$

$$= \frac{\rho_1 c_1}{\rho_3 c_3} \frac{1}{\frac{1}{4}\left(1 + \frac{\rho_1 c_1}{\rho_3 c_3}\right)^2 \cos^2 k_2 L + \frac{1}{4}\left(\frac{\rho_2 c_2}{\rho_3 c_3} + \frac{\rho_1 c_1}{\rho_2 c_2}\right)^2 \sin^2 k_2 L} \tag{4.39}$$

For calculating the transmission of a wave through three layers it was necessary to define five different waves and to carry out some lengthy algebra in order to obtain the desired result. If more layers were to be considered then the algebra would become prohibitive. It is possible, however, to write down the pressure and particle velocity in matrix form and obtain a characteristic matrix for each bounded layer except the two outside ones. The characteristic matrices are then multiplied together to give just one matrix for the complete layered system. The same boundary conditions of continuity of pressure and continuity of particle velocity are still applied between layers and the overall ratios of transmitted or reflected pressure to incident pressure can be extracted after the characteristic matrices have been combined. The method will be demonstrated analytically for a three layer system but, in matrix form, it is really more suited to numerical calculations using a computer.

If we take the three layers illustrated in Fig. 4.3 as our example, it is the middle bounded layer for which we set up a characteristic matrix. The total pressure and particle velocity at any point in medium 2 are given by

$$\mathbf{p} = A_2\, e^{j(\omega t - k_2 x)} + B_2\, e^{j(\omega t + k_2 x)}$$

$$\mathbf{u} = \frac{A_2}{\rho_2 c_2}\, e^{j(\omega t - k_2 x)} - \frac{B_2}{\rho_2 c_2}\, e^{j(\omega t + k_2 x)}$$

Written in matrix form these become

$$\begin{bmatrix} \mathbf{p} \\ \\ \mathbf{u} \end{bmatrix} = \begin{bmatrix} e^{-jk_2 x} & e^{jk_2 x} \\ \dfrac{e^{-jk_2 x}}{\rho_2 c_2} & -\dfrac{e^{jk_2 x}}{\rho_2 c_2} \end{bmatrix} \times \begin{bmatrix} A_2\, e^{j\omega t} \\ \\ B_2\, e^{j\omega t} \end{bmatrix}$$

Pressure and particle velocity are continuous at the boundaries and so we find their specific values for this layer by substituting the values $x = 0$ and $x = L$ into the general matrix equation.
i.e. at $x = 0$

$$\begin{bmatrix} p_0 \\ u_0 \end{bmatrix} = \begin{bmatrix} 1 & 1 \\ \dfrac{1}{\rho_2 c_2} & -\dfrac{1}{\rho_2 c_2} \end{bmatrix} \times \begin{bmatrix} A_2\, e^{j\omega t} \\ B_2\, e^{j\omega t} \end{bmatrix} \tag{4.40}$$

and at $x = L$

$$\begin{bmatrix} p_L \\ u_L \end{bmatrix} = \begin{bmatrix} e^{-jk_2 L} & e^{jk_2 L} \\ \dfrac{e^{-jk_2 L}}{\rho_2 c_2} & -\dfrac{e^{jk_2 L}}{\rho_2 c_2} \end{bmatrix} \times \begin{bmatrix} A_2\, e^{j\omega t} \\ B_2\, e^{j\omega t} \end{bmatrix} \tag{4.41}$$

The matrix containing $A_2\, e^{j\omega t}$ and $B_2\, e^{j\omega t}$ is common to both equations (4.40) and (4.41) and by eliminating this term the pressure and particle velocity at $x = 0$ is obtained in terms of those at $x = L$.
From equation (4.41)

$$\begin{bmatrix} A_2\, e^{j\omega t} \\ B_2\, e^{j\omega t} \end{bmatrix} = \begin{bmatrix} e^{-jk_2 L} & e^{jk_2 L} \\ \dfrac{e^{-jk_2 L}}{\rho_2 c_2} & -\dfrac{e^{jk_2 L}}{\rho_2 c_2} \end{bmatrix}^{-1} \times \begin{bmatrix} p_L \\ u_L \end{bmatrix}$$

and substitution into equation (4.40) gives

$$\begin{bmatrix} p_0 \\ u_0 \end{bmatrix} = \begin{bmatrix} 1 & 1 \\ \dfrac{1}{\rho_2 c_2} & -\dfrac{1}{\rho_2 c_2} \end{bmatrix} \times \begin{bmatrix} e^{-jk_2 L} & e^{jk_2 L} \\ \dfrac{e^{-jk_2 L}}{\rho_2 c_2} & -\dfrac{e^{jk_2 L}}{\rho_2 c_2} \end{bmatrix}^{-1} \times \begin{bmatrix} p_L \\ u_L \end{bmatrix}$$

$$= \begin{bmatrix} \cos k_2 L & j\rho_2 c_2 \sin k_2 L \\ \dfrac{j \sin k_2 L}{\rho_2 c_2} & \cos k_2 L \end{bmatrix} \times \begin{bmatrix} p_L \\ u_L \end{bmatrix} \tag{4.42}$$

The matrix relating p_0 and u_0 to p_L and u_L in equation (4.42) is called the characteristic matrix of the layer and comprises terms made up from the physical properties of the layer. There is, of course, no need to repeat the algebra because every layer has a characteristic matrix of the same form. If there were more than one bounded layer then the individual layer characteristic matrices would be multiplied together to give one matrix for the multiple layer system. In this case equation (4.42) is complete and it is now necessary to write $\mathbf{p}_0$ and $\mathbf{u}_0$ in terms of their values on the medium 1 side of the boundary, *i.e.* in terms of $\mathbf{A}_1$ and $\mathbf{B}_1$, and $\mathbf{p}_L$ and $\mathbf{u}_L$ in terms of their values on the medium 3 side of the boundary, *i.e.* in terms of $\mathbf{A}_3$. So

$$
\begin{bmatrix} \mathbf{A}_1\, e^{j\omega t} + \mathbf{B}_1\, e^{j\omega t} \\[2pt] \dfrac{\mathbf{A}_1\, e^{j\omega t} - \mathbf{B}_1\, e^{j\omega t}}{\rho_1 c_1} \end{bmatrix} = \begin{bmatrix} \cos k_2 L & j\rho_2 c_2 \sin k_2 L \\[4pt] \dfrac{j \sin k_2 L}{\rho_2 c_2} & \cos k_2 L \end{bmatrix} \times \begin{bmatrix} \mathbf{A}_3\, e^{j\omega t} \\[2pt] \dfrac{\mathbf{A}_3\, e^{j\omega t}}{\rho_3 c_3} \end{bmatrix}
$$

Dividing through by $e^{j\omega t}$ and taking $\rho_1 c_1$ and $\rho_3 c_3$ into the central matrix

$$
\begin{bmatrix} \mathbf{A}_1 + \mathbf{B}_1 \\[6pt] \mathbf{A}_1 - \mathbf{B}_1 \end{bmatrix} = \begin{bmatrix} \cos k_2 L & j\,\dfrac{\rho_2 c_2}{\rho_3 c_3} \sin k_2 L \\[8pt] j\,\dfrac{\rho_1 c_1}{\rho_2 c_2} \sin k_2 L & \dfrac{\rho_1 c_1}{\rho_3 c_3} \cos k_2 L \end{bmatrix} \times \begin{bmatrix} \mathbf{A}_3 \\[6pt] \mathbf{A}_3 \end{bmatrix}
$$

The matrices can now be thought of as two equations again and $\mathbf{B}_1$ is eliminated by addition, leaving

$$
\mathbf{A}_1 = \frac{\mathbf{A}_3}{2} \left\{ \left(1 + \frac{\rho_1 c_1}{\rho_3 c_3}\right) \cos k_2 L + j \left(\frac{\rho_2 c_2}{\rho_3 c_3} + \frac{\rho_1 c_1}{\rho_2 c_2}\right) \sin k_2 L \right\} \qquad (4.43)
$$

Equation (4.43) is identical to (4.38) as one would expect since basically the same operations have been carried out. The matrix method has been used extensively for calculating transmission through layered optical media where the wave equations are analogous to the acoustical case and it is simple to extend its use in acoustics to transmission at any angle of incidence through fluid layers. To extend it to transmission through solid layers is more difficult because then shear waves must also be included.

4.5 Special Case of Transmission Through a Wall

A very common case of sound transmission is through a wall separating two rooms. Both medium 1 and medium 3 are air. Putting $\rho_3 c_3 = \rho_1 c_1$ in equation (4.39) gives

$$a_t = \frac{1}{\cos^2 k_2 L + \frac{1}{4}(\rho_2 c_2/\rho_1 c_1 + \rho_1 c_1/\rho_2 c_2)^2 \sin^2 k_2 L}$$

Medium 2 is a solid and normally $\rho_2 c_2 \gg \rho_1 c_1$.

therefore
$$a_t \simeq \frac{1}{\cos^2 k_2 L + \frac{1}{4}(\rho_2 c_2/\rho_1 c_1)^2 \sin^2 k_2 L}$$

Except at very high frequencies, $k_2 L$ is small and

$$\sin k_2 L \simeq k_2 L = \omega L/c_2; \quad \cos k_2 L \simeq 1.$$

therefore
$$a_t \simeq \frac{1}{1 + \frac{1}{4}(\rho_2 c_2/\rho_1 c_1)^2 (\omega L/c_2)^2}$$

$$= \frac{1}{1 + (\omega M/2\rho_1 c_1)^2} \tag{4.44}$$

where $M (= \rho_2 L)$ is the mass per unit area of the wall.

It is common practice to quote the insulation of a wall in the logarithmic (decibel) form

i.e. Sound reduction $= 10 \log_{10} (1/a_t)$ dB

$$= 10 \log_{10} \left[1 + \left(\frac{\omega M}{2\rho_1 c_1} \right)^2 \right] \text{dB} \tag{4.45}$$

Equation (4.45) has been derived for normal incidence sound waves and since the insulation depends upon the mass per unit area of the wall it is often referred to as the normal incidence mass law. In practice sound waves strike the wall at all angles of incidence but a complete analysis is beyond the scope of this book. It is sufficient to note that the transmission coefficient increases as the angle of incidence departs from the normal, and the field insulation, which includes all angles of incidence met with in reality, is given by

$$(SR)_{\text{FIELD}} = 10 \log_{10} \left[1 + \left(\frac{\omega M}{2\rho_1 c_1} \right)^2 \right] - 5 \text{ dB} \tag{4.46}$$

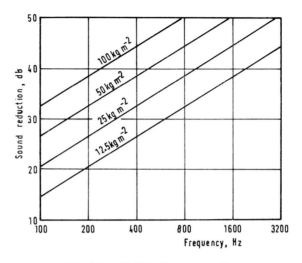

Fig. 4.4.   Field incidence mass law.

The sound reduction predicted by equation (4.46) is displayed as a function of frequency and of mass/unit area of the wall in Fig. 4.4. So, if the wall obeys the mass law, the insulation is increased by 6 dB by doubling either the frequency or the mass per unit area.

CHAPTER 5

Transmission Through Ducts

5.1 Acoustic Impedance

In this chapter we are concerned with one-dimensional sound waves travelling through ducts. The medium does not change but the cross-sectional area may or there may be a side branch and whenever a discontinuity occurs, some of the incident wave is reflected. To determine the magnitudes of the transmitted and reflected waves, we may apply certain conditions which the waves must satisfy. The first condition is continuity of pressure, the second is continuity of volume velocity, where volume velocity is particle velocity times the cross-sectional area. Combining these leads to the concept of acoustic impedance which is defined as

$$\mathbf{Z} = \frac{\text{pressure}}{\text{volume velocity}} = \frac{\mathbf{p}}{\mathbf{U}} = \frac{\mathbf{p}}{S \cdot \mathbf{u}} = \frac{\mathbf{z}}{S} \tag{5.1}$$

where S is the cross-sectional area.

Because of the dependence on cross-sectional area, total acoustic power has more significance than acoustic intensity. The power associated with a plane wave is

$$W_+ = I_+ \times S = \frac{\bar{p}^2}{\rho_0 c} \times S = \frac{\bar{p}^2}{\mathbf{Z}_0} \tag{5.2}$$

and again it is convenient to use acoustic impedance.

5.2 The Helmholtz Resonator

The Helmholtz resonator is of interest in its own right because it has been used in acoustics for at least 2000 years. Its application will be discussed in the section on absorbers in Chapter 6. It is of additional interest here in demonstrating the use of acoustic impedance and in introducing the concepts of acoustical stiffness and mass. In its simplest form it consists of an enclosed volume, V, connected to the main body of fluid by a small circular aperture of length l and area $S(= \pi a^2)$. It is assumed that all dimensions are much less than the wavelength of sound, which restricts its use to low frequencies. The resonator is now exactly the same as a mechanical mass-spring-damper system with the plug of fluid in the

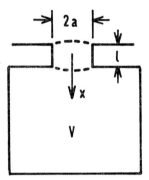

Fig. 5.1.   *Representation of a Helmholtz resonator.*

neck acting as the mass and the enclosed volume of fluid acting as a spring. The frictional forces acting on the plug of fluid in the neck are negligible unless the aperture is very small or is intentionally restricted in some way. However, as the plug vibrates it can radiate sound into the main body of fluid and this energy is lost to the resonator. In the absence of restrictions in the neck it is this radiation term which must be taken into account. To set up the equation of motion the plug of fluid is assumed to be displaced inwards by an amount x by an external force and the four terms are obtained as follows.

(1) *Stiffness of enclosed volume*

The oscillations are sufficiently rapid for the expansions and contractions to be adiabatic. Therefore we may use equation (2.27), putting $\delta V = -S\mathbf{x}$

i.e. $$\mathbf{p} = \frac{\gamma P_0 S}{V}\mathbf{x} = \rho_0 c^2 \frac{S}{V}\mathbf{x}$$

The force in the $+x$ direction $= -\mathbf{p}S = -\rho_0 c^2(S^2/V)\mathbf{x}$ (5.3)

(2) *Mass of the plug of fluid*

The plug of moving fluid is larger than the actual volume, Sl, and is usually written as the effective volume $Sl' = S(l+2\Delta l)$. $S\Delta l$ is the same as the volume of air which moves with a circular piston of area S vibrating in an infinite baffle and it appears twice because the plug has two ends.

Therefore the total mass of fluid

$$= \rho_0 Sl'$$

$$= \rho_0 S(l+2\Delta l) \tag{5.4}$$

(3) *Radiation resistance force*

The radiation from the plug of air is the same as the radiation from a circular piston in a rigid baffle, but it is convenient for the moment merely to write it in the form of a constant R.

i.e. $$\mathbf{F}_{\text{RAD}} = -RS^2 \frac{d\mathbf{x}}{dt} \tag{5.5}$$

(4) *Driving force*

If the resonator is driven by a harmonic sound wave then

$$\mathbf{F} = S\mathbf{p}$$

$$= SA\, e^{j\omega t} \tag{5.6}$$

Setting up the equation of motion gives

$$\rho_0 l'S \frac{d^2\mathbf{x}}{dt^2} + RS^2\frac{d\mathbf{x}}{dt} + \frac{\rho_0 c^2}{V}S^2\mathbf{x} = SA\, e^{j\omega t} \tag{5.7}$$

The particle velocity

$$\mathbf{u} = \frac{d\mathbf{x}}{dt}$$

and, therefore the volume velocity

$$\mathbf{U} = S\frac{d\mathbf{x}}{dt}$$

therefore $\qquad \dfrac{\rho_0 l'}{S}\dfrac{d\mathbf{U}}{dt} + R\mathbf{U} + \dfrac{\rho_0 c^2}{V}\int \mathbf{U}\,dt = \mathbf{A}\,e^{j\omega t}$

or $\qquad\qquad M\dfrac{d\mathbf{U}}{dt} + R\mathbf{U} + K\int \mathbf{U}\,dt = \mathbf{A}\,e^{j\omega t} \qquad\qquad (5.8)$

where $\quad M = \rho_0 l'/S \quad$ is the acoustic mass

and $\qquad K = \rho_0 c^2/V \quad$ is the acoustic stiffness $\qquad (5.9)$

The harmonic solution of equation (5.8) is

$$\mathbf{U} = \frac{\mathbf{A}\,e^{j\omega t}}{R + j(\omega M - K/\omega)} = \frac{\mathbf{p}}{\mathbf{Z}} \qquad\qquad (5.10)$$

Resonance is defined as occurring when $\mathbf{U}$ reaches a maximum; that is when the imaginary or reactive part of the acoustic impedance is zero

i.e. $\qquad\qquad \omega_0^2 = \dfrac{K}{M} = \dfrac{\rho_0 c^2/V}{\rho_0 l'/S} = \dfrac{c^2 S}{l'V}$

therefore $\qquad f_0 = \dfrac{c}{2\pi}\left(\dfrac{S}{Vl'}\right)^{1/2} = \dfrac{c}{2\pi}\left(\dfrac{\pi a^2}{V(l+2\Delta l)}\right)^{1/2} \qquad (5.11)$

At the resonant frequency the response of the resonator is controlled entirely by the damping in the system, in this case coming from the radiation term, and, if the damping is small, large acoustic pressures can be produced inside the resonator. At other frequencies the response is less and the acoustic pressures inside the resonator are less. The resonator could therefore be regarded as a tuned filter, with the centre frequency determined by its dimensions and the band width determined by the damping.

Equations (5.9) are important in that they give the general form for an acoustic mass and an acoustic stiffness. These will be used again later when considering lumped acoustic elements.

Some useful conclusions can be drawn from the acoustic impedance of the resonator which may be re-written

$$\mathbf{Z} = R + j(\omega M - K/\omega)$$

$$= R + j\left\{\frac{\omega \rho_0}{S}\,(l + 2\Delta l) - \frac{\rho_0 c^2}{\omega V}\right\} \tag{5.12}$$

First of all, the acoustic impedance of an acoustic mass and an acoustic stiffness are given by $j\omega M$ and $-jK/\omega$ respectively. Secondly, we could deduce that the acoustic impedance of a circular piston of radius a, where $a \ll \lambda$, vibrating in a rigid infinite baffle should consist of the radiation term R and a mass term corresponding to one end correction

i.e. $$\mathbf{Z}_{\text{piston}} = R + \frac{j\omega \rho_0}{S}\Delta l \tag{5.13}$$

But the radiation impedance of such a piston has already been established in Chapter 3 and is given by equation (3.29). The radiation impedance and the acoustic impedance differ, by definition, by a factor of S^2 and comparison shows that

$$\mathbf{Z}_{\text{piston}} = \frac{\rho_0 c k^2}{2\pi} + \frac{j\omega \rho_0}{S}\cdot\frac{8a}{3\pi} \tag{5.14}$$

The end correction, $\Delta l = 8a/3\pi$, was quoted in Chapter 2 in an example involving sound waves in an open-ended pipe.

5.3 The Reflection of Waves in a Pipe

If a single wave is travelling in the positive x direction along a tube of constant cross-sectional area, S, the acoustic impedance seen by that wave will be $\rho_0 c/S$. Suppose now that there is some discontinuity in the pipe, *e.g.* a change in cross-sectional area, then, in general, some of the incident wave will be reflected and the acoustic impedance of the reflected wave will be $-\rho_0 c/S$.

The pressures and volume velocities of the two individual waves are

$$\mathbf{p}_i = \mathbf{A}\,e^{j(\omega t - kx)}; \quad \mathbf{U}_i = \frac{\mathbf{p}_i}{\rho_0 c / S} \tag{5.15}$$

$$\mathbf{p}_r = \mathbf{B}\,e^{j(\omega t + kx)}; \quad \mathbf{U}_r = -\frac{\mathbf{p}_r}{\rho_0 c / S} \tag{5.16}$$

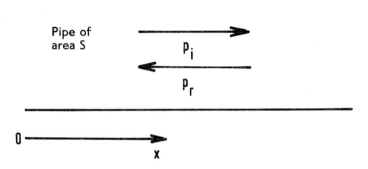

Fig. 5.2.   *Incident and reflected waves in a pipe.*

With both waves present the total acoustic impedance at any point x is given by

$$\mathbf{Z}_x = \frac{\mathbf{p}_i + \mathbf{p}_r}{\mathbf{U}_i + \mathbf{U}_r} = \frac{\rho_0 c}{S} \frac{\mathbf{p}_i + \mathbf{p}_r}{\mathbf{p}_i - \mathbf{p}_r}$$

$$= \frac{\rho_0 c}{S} \frac{\mathbf{A}\,e^{-jkx} + \mathbf{B}\,e^{jkx}}{\mathbf{A}\,e^{-jkx} - \mathbf{B}\,e^{jkx}} \tag{5.17}$$

Because of the requirements of continuity of pressure and continuity of volume velocity there must be continuity of acoustic impedance at any point in the pipe. Therefore, if at some point the acoustic impedance is known to have a particular value, then $\mathbf{Z}_x$ given by equation (5.17) must also take that value. For convenience, let us assume that the acoustic impedance is known to be $\mathbf{Z}_0$ at $x = 0$.

Then

$$\mathbf{Z}_0 = \frac{\rho_0 c}{S} \frac{\mathbf{A} + \mathbf{B}}{\mathbf{A} - \mathbf{B}} \tag{5.18}$$

This may be re-arranged to give

$$\frac{\mathbf{B}}{\mathbf{A}} = \frac{\mathbf{Z}_0 - \rho_0 c/S}{\mathbf{Z}_0 + \rho_0 c/S} \qquad (5.19)$$

Equation (5.19) may be applied to the case of a sudden change in cross-sectional area of the pipe, as is illustrated in Fig. 5.3.

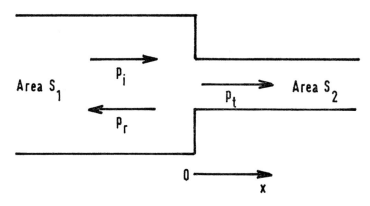

Fig. 5.3 *Change in cross-sectional area in a pipe.*

Provided that only the single wave $\mathbf{p}_t$ exists in the pipe of area S_2, *i.e.* there are no more reflections farther along the pipe, the acoustic impedance at the origin is $\rho_0 c/S_2$.

Therefore
$$\frac{\mathbf{B}}{\mathbf{A}} = \frac{\rho_0 c/S_2 - \rho_0 c/S_1}{\rho_0 c/S_2 + \rho_0 c/S_1}$$

$$= \frac{S_1 - S_2}{S_1 + S_2} \qquad (5.20)$$

In order to calculate how much of the incident energy is reflected and how much is transmitted, we may write the power reflection and transmission coefficients as follows.

$$\alpha_r = \frac{W_r}{W_i} = \frac{|\mathbf{B}|^2/2\rho_0 c/S_1}{|\mathbf{A}|^2/2\rho_0 c/S_1} = \left|\frac{\mathbf{B}}{\mathbf{A}}\right|^2 \qquad (5.21)$$

$$= \frac{(S_1 - S_2)^2}{(S_1 + S_2)^2}$$

$$a_t = 1 - a_r = \frac{4S_1 S_2}{(S_1 + S_2)^2} \tag{5.22}$$

So a_r and a_t are dependent only on the ratio $S_1 : S_2$ or $S_2 : S_1$ and not on whether the cross-sectional area decreases or increases. Furthermore the ratio has to be quite large to produce a substantial reduction in transmitted energy. For example, if $S_2 : S_1 = 10$ (or $0 \cdot 1$), $a_t = 0 \cdot 33$.

5.4 Resonance in Pipes

The concept of acoustic impedance can easily be used to investigate the behaviour of pipes of finite length. Consider a pipe of uniform cross-sectional area $S(= \pi a^2)$ and of length L. Suppose that the acoustic impedance is known to be $\mathbf{Z}_0$ at $x = 0$ and $\mathbf{Z}_L$ at $x = L$. Then, from equation (5.17)

$$\mathbf{Z}_L = \frac{\rho_0 c}{S} \frac{\mathbf{A} e^{-jkL} + \mathbf{B} e^{jkL}}{\mathbf{A} e^{-jkL} - \mathbf{B} e^{jkL}}$$

and

$$\mathbf{Z}_0 = \frac{\rho_0 c}{S} \frac{\mathbf{A} + \mathbf{B}}{\mathbf{A} - \mathbf{B}}$$

Combining these two equations to eliminate $\mathbf{A}$ and $\mathbf{B}$ gives

$$\mathbf{Z}_0 = \frac{\rho_0 c}{S} \frac{\mathbf{Z}_L + j(\rho_0 c/S) \tan kL}{(\rho_0 c/S) + j\mathbf{Z}_L \tan kL} \tag{5.23}$$

If we consider a pipe first with a closed end and then with an open end, we shall be able to obtain the conditions for resonance in the pipe.

(1) *Closed end*

$$\mathbf{Z}_L = \infty$$

therefore

$$\mathbf{Z}_0 = \frac{\rho_0 c}{S} \frac{1}{j \tan kL}$$

The condition for resonance is that the reactive or imaginary part of $\mathbf{Z}_0$ should be zero. This implies that the pipe is open and can be driven at the origin.

Then $$\tan kL = \infty$$

and $$kL = \frac{(2n-1)\pi}{2}$$

(2) *Open end*

From equation (5.14)

$$\mathbf{Z}_L = \frac{\rho_0 c}{S}\left(\frac{k^2 a^2}{2} + j\frac{8ka}{3\pi}\right)$$

If $ka \ll 1$ then, numerically, the real part of this impedance is negligible compared with the imaginary part and so

$$\mathbf{Z}_L \simeq j\frac{\rho_0 c}{S}\frac{8ka}{3\pi}$$

therefore $$\mathbf{Z}_0 = \frac{\rho_0 c}{S}\frac{j(8ka/3\pi)+j\tan kL}{1-(8ka/3\pi)\tan kL}$$

and the imaginary part is zero when

$$\frac{8ka}{3\pi}+\tan kL = 0$$

Since $8ka/3\pi$ is very small then $\tan kL \simeq kL-n\pi$ and

$$8ka/3\pi+kL = n\pi$$

i.e. $$k = \frac{n\pi}{(L+8a/3\pi)}$$

or $$f = \frac{nc}{2(L+8a/3\pi)}$$

where $8a/3\pi$ is the end correction for a flanged pipe.

5.5   The Effect of a Side Branch

The main pipe, of area S, has a side branch of acoustic impedance $\mathbf{Z}_b$ at the position $x = 0$. There are no reflected waves beyond the side branch and the acoustic impedance of the main pipe at $x = 0$

is $\rho_0 c/S$. The side branch causes a reflection of the incident wave and, at $x = 0$, we may write the four pressures and volume velocities as

$$p_i = A_1 e^{j\omega t} \qquad U_i = \frac{p_i}{\rho_0 c/S} \tag{5.24}$$

$$p_r = B_1 e^{j\omega t} \qquad U_r = -\frac{p_r}{\rho_0 c/S} \tag{5.25}$$

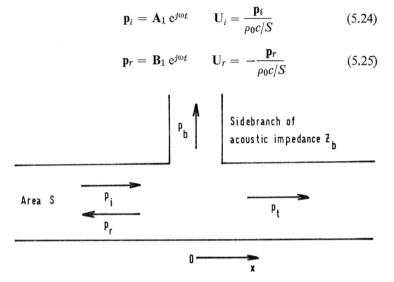

Fig. 5.4. *Waves in a pipe with a side branch.*

$$p_t = A_2 e^{j\omega t} \qquad U_t = \frac{p_t}{\rho_0 c/S} \tag{5.26}$$

$$p_b = A_3 e^{j\omega t} \qquad U_b = \frac{p_b}{Z_b} \tag{5.27}$$

We must consider the conditions of pressure and volume velocity separately.

(1) *Continuity of pressure*

$$p_i + p_r = p_t = p_b \tag{5.28}$$

(2) *Continuity of volume velocity*

$$U_c + U_r = U_t + U_b \tag{5.29}$$

Combining equations (5.26) and (5.27) we obtain

$$\frac{U_c + U_r}{p_i + p_r} = \frac{U_t}{p_t} + \frac{U_b}{p_b}$$

or

$$\frac{1}{Z} = \frac{1}{Z_t} + \frac{1}{Z_b} \tag{5.30}$$

where

$$Z = \frac{\rho_0 c}{S} \frac{A_1 + B_1}{A_1 - B_1} \quad \text{and} \quad Z_t = \frac{\rho_0 c}{S}$$

So in this instance we find that the acoustic impedances of the two branches add together in the same way as two electrical impedances which are in parallel.

Equation (5.30) is easily re-arranged to give

$$\frac{B_1}{A_1} = \frac{-\rho_0 c/2S}{\rho_0 c/2S + Z_b} \tag{5.31}$$

Then from equation (5.29)

$$\frac{A_2}{A_1} = \frac{A_3}{A_1} = \frac{Z_b}{\rho_0 c/2S + Z_b} \tag{5.32}$$

In general the acoustic impedance of the side branch is complex and may be written

$$Z_b = R_b + jX_b \tag{5.33}$$

Then the power coefficients are as follows

$$a_r = \left|\frac{B_1}{A_1}\right|^2 = \frac{(\rho_0 c/2S)^2}{(\rho_0 c/2S + R_b)^2 + X_b^2} \tag{5.34}$$

$$a_t = \left|\frac{A_2}{A_1}\right|^2 = \frac{R_b^2 + X_b^2}{(\rho_0 c/2S + R_b)^2 + X_b^2} \tag{5.35}$$

$$a_b = 1 - a_r - a_t = \frac{R_b \cdot \rho_0 c/2S}{(\rho_0 c/2S + R_b)^2 + X_b^2} \tag{5.36}$$

The transmitted power, a_t, can be zero when both R_b and X_b are zero. Under these circumstances a_b is also zero and so the power is totally reflected back to the source.

5.6   A Helmholtz Resonator as a Side Branch

The acoustic impedance of a Helmholtz resonator, given in equation (5.12) included a radiation term. When used as a side branch the radiated energy is returned to the main pipe and so the impedance of the resonator is

$$\mathbf{Z}_b = j \left(\frac{\omega \rho_0 l'}{S_b} - \frac{\rho_0 c^2}{\omega V} \right) \quad \text{where } l' = l + 16a/3\pi$$

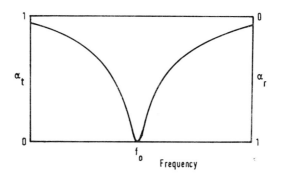

Fig. 5.5.   Power coefficients with a Helmholtz resonator as a side branch.

Substituting into equation (5.35) gives

$$a_t = \frac{1}{1 + \dfrac{(\rho_0 c/2S)^2}{[(\omega \rho_0 l'/S_b) - (\rho_0 c^2/\omega V)]^2}} = \frac{1}{1 + \dfrac{c^2}{4S^2[(\omega l'/S_b) - (c^2/\omega V)]^2}}$$

$a_t = 0$ at the resonant frequency of the resonator, *i.e.* when $f = f_0 = (c/2\pi)(S_b/l'V)^{1/2}$, a_b is always zero. Fig. 5.5 illustrates a_t and a_r as a function of frequency.

A tube with a closed end used as a side branch would have a similar effect to the Helmholtz resonator. The acoustic impedance is zero at all of the resonant frequencies of the tube and the transmitted power is also zero. It is sometimes referred to as a quarter-wave tube since the fundamental frequency is such that the length of the tube equals one quarter of a wavelength.

5.7 An Orifice as a Side Branch

The impedance of an orifice is composed of two terms; the radiation external to the orifice and the inertia of the gas in the orifice. It is in fact the same as for a Helmholtz resonator of zero stiffness.

i.e. $$Z_b = \frac{\rho_0 c}{S} \cdot \frac{k^2 a^2}{2} + j \frac{\omega \rho_0}{S} l' \quad \text{where} \quad l' = l + 16a/3\pi$$

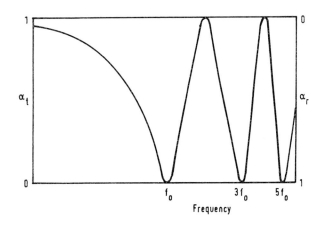

Fig. 5.6.   Power coefficients with a quarter-wave tube as a side branch.

Before substituting this impedance directly into the power coefficient equations it is worth comparing the magnitudes of the real and imaginary components.

i.e. $$\frac{R_b}{X_b} = \frac{(\rho_0 c/S)(k^2 a^2/2)}{(\omega \rho_0/S)(l + 16a/3\pi)} = \frac{ka^2}{2(l + 16a/3\pi)}$$

The limiting case is for a thin walled pipe when $l \simeq 0$ and

$$\frac{R_b}{X_b} \leqslant \frac{3\pi}{32} ka$$

Provided that the orifice is small compared with the wavelength then $ka \ll 1$ and R_b is negligible compared with X_b. The transmission coefficient is given by

$$a_t = \frac{1}{1+\{[\rho_0 c/2S]/[(\omega\rho_0/S)l']\}^2} = \frac{1}{1+(c/2\omega l')^2}$$

At low frequencies a_t is small but as ω increases so a_t tends to unity and we have a high-pass filter. This is illustrated in Fig. 5.7.

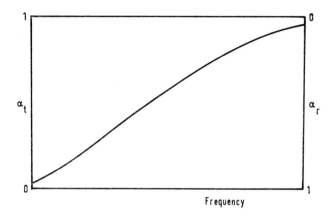

Fig. 5.7. Power coefficients with an orifice as a side branch.

It has been assumed that R_b is negligible and consequently a_b is also negligible. Therefore any power which is not transmitted is reflected back to the source.

5.8   The Effect of an Expansion Chamber

Provided that the dimensions of the expansion chamber are small compared to the wavelength of sound it may be considered as a side branch. There is no means of dissipating sound energy and R_b is zero. The volume of the chamber provides stiffness and the acoustic impedance of the side branch may be written

$$\mathbf{Z}_b = -j\frac{\rho_0 c^2}{\omega S_2 L}$$

Substituting into equation (5.35) to obtain the transmission coefficient

$$a_t = \frac{1}{1 + [(\rho_0 c/2S_1)(\omega S_2 L/\rho_0 c^2)]^2} = \frac{1}{1 + (\omega S_2 L/2S_1 c)^2}$$

At low frequencies a_t is very nearly unity but as the frequency increases so a_t decreases, and the system acts as a low-pass filter. a_b is zero and the energy which is not transmitted is reflected to the source. This is illustrated in Fig. 5.9.

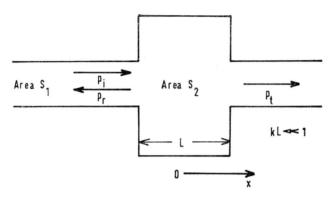

Fig. 5.8. An expansion chamber in a pipe.

In fact a_t does not steadily decrease as the frequency increases because the condition $kL \ll 1$ ceases to hold. For every frequency that the length L equals an integral number of half-wavelengths the sound is easily transmitted as is indicated by the broken line in Fig. 5.9.

The more practical case is when the tailpipe attached to the expansion chamber is of a finite length rather than infinite as has been assumed. Then at low frequencies the tailpipe and the expansion chamber behave rather like a Helmholtz resonator, that is the chamber provides stiffness and the tailpipe, of effective length l', provides mass. The fundamental resonance of the system is given by

$$f_0 = \frac{c}{2\pi} \left(\frac{S_1}{S_2 L l'} \right)^{1/2}$$

At resonance the sound is transmitted easily and a peak appears in the power transmission coefficient curve at around f_0. This is depicted in Fig. 5.10. Normally this frequency would be kept as low as possible either by making the expansion chamber large or the tailpipe long and thin.

At higher frequencies, the resonances corresponding to standing waves along the expansion chamber still occur but now extra

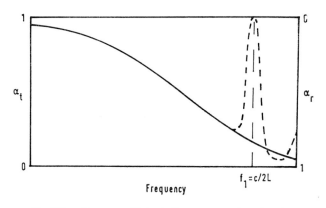

Fig. 5.9. *Power coefficients for an expansion chamber.*

resonances are introduced whenever the length of the tailpipe corresponds to an integral number of half-wavelengths, *i.e.* whenever $f = nc/2l'$, and each of these produces a small peak in the transmission coefficient curve.

5.9 Sound Waves in Horns

So far in this chapter a reflected wave has always resulted from a discontinuity in the pipe. Sometimes it is necessary to change the cross-sectional area without wanting a reflected wave and to achieve this a horn section is used. Horns are also very useful for improving the performance of loudspeakers at low frequencies because they provide a means of matching the acoustic impedance of the space to that of the speaker. The one-dimensional wave equation derived in Chapter 2 is not appropriate because it was assumed there that the cross-sectional area was constant. The exact wave equation is very difficult to set up for the majority of horn shapes but fortunately its use is rarely justified. If we make the following simplifying assumptions it is possible to set up an approximate wave equation which is adequate for our purposes.

1. Second order terms may be neglected.
2. Waves move parallel to the axis.
3. The walls are all perfectly rigid.
4. The waves remain in contact with the walls.

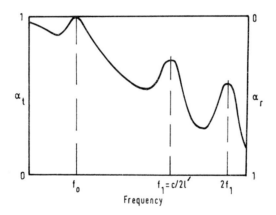

Fig. 5.10. Power coefficients for an expansion chamber with a finite length tailpipe.

Referring to the variable cross-section shown in Fig. 5.11, we must suppose that an element of fluid at x, of volume $S\, \delta x$, is disturbed by the passage of a sound wave to a position $x + \xi$. Its new cross-sectional area is $S + \xi(\partial S/\partial x)$ and its length is $\delta x + \delta \xi$.

Therefore the increase in volume, neglecting second order terms, is

$$\delta V = \left(S + \xi \frac{\partial S}{\partial x}\right)(\delta x + \delta \xi) - S\, \delta x$$

$$\simeq S\, \delta \xi + \xi \frac{\partial S}{\partial x}\, \delta x$$

and
$$\frac{\delta V}{V_0} = \frac{\partial \xi}{\partial x} + \xi \frac{1}{S}\frac{\partial S}{\partial x}$$

$$= \frac{1}{S}\frac{\partial(S\xi)}{\partial x}$$

Equation (2.27) derived from the adiabatic gas law still applies and so

$$p = -\gamma P_0 \frac{1}{S}\frac{\partial(S\xi)}{\partial x} \qquad (5.37)$$

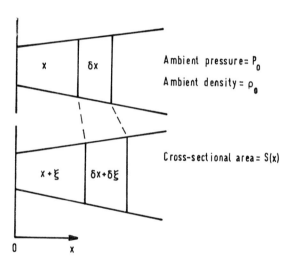

Fig. 5.11.   *Displacement of an element of gas in a horn due to the passage of a longitudinal wave.*

which is to be compared with equation (2.28). To a first order approximation equation (2.26) also remains unaltered and the wave equation is

$$\frac{1}{S}\frac{\partial}{\partial x}\left(S\frac{\partial p}{\partial x}\right) = \frac{1}{c^2}\frac{\partial^2 p}{\partial t^2} \qquad (5.38)$$

The velocity c is the same as before and the particle velocity is still related to the pressure by equation (2.33).

5.10 The Exponential Horn

Although we could equally well consider the conical or catenoidal horn, one example will suffice and the exponential horn is probably of most general interest. If the horn flares from a throat area of S_0 at the origin then the area at any point is given by

$$S = S_0\, e^{mx} \qquad (5.39)$$

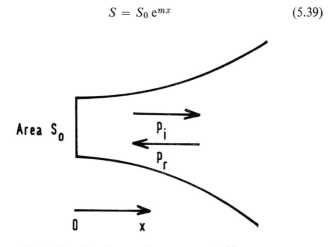

Fig. 5.12. *Sound waves in an exponential horn.*

where m is known as the flare constant.
Substituting this expression for S into equation (5.38) gives

$$\frac{\partial^2 p}{\partial x^2} + m\,\frac{\partial p}{\partial x} = \frac{1}{c^2}\,\frac{\partial^2 p}{\partial t^2} \qquad (5.40)$$

We may assume a solution of the general form

$$\mathbf{p} = \mathbf{A}\, e^{j(\omega t + \gamma x)}$$

and substitution of this solution into the wave equation yields the requirement that

$$\gamma^2 - jm\gamma - k^2 = 0 \quad \text{where} \quad k = \omega/c$$

therefore $$\gamma = j\frac{m}{2} \pm \left(k^2 - \frac{m^2}{4}\right)^{1/2} = ja \pm \beta$$

and the solution for **p** may be written

$$\mathbf{p} = (\mathbf{A}\ e^{j(\omega t - \beta x)} + \mathbf{B}\ e^{j(\omega t + \beta x)})\ e^{-\alpha x} \tag{5.41}$$

This represents two waves, one travelling in each direction, which decrease exponentially with increasing x and which have a phase velocity given by

$$c' = \frac{\omega}{\beta}\frac{c}{[1 - (m^2/4k^2)]^{1/2}}$$

The phase velocity may be thought of as the velocity at which a point of maximum pressure travels through the horn and at very high frequencies it tends towards c, the normal velocity of sound. As the frequency reduces, however, the phase velocity increases until at a frequency corresponding to $k_c = m/2$ it becomes infinite. Below this frequency the phase velocity is imaginary and waves are not propagated. The particular frequency is known as the cutoff frequency and is given by

$$f_c = \frac{mc}{4\pi} \tag{5.42}$$

The particle velocity corresponding to equation (5.41) is

$$\mathbf{u} = -\frac{1}{\omega\rho_0}[(j a - \beta)\mathbf{A}\ e^{j(\omega t - \beta x)} + (j a + \beta)\mathbf{B}\ e^{j(\omega t + \beta x)}]\ e^{-\alpha x}$$

and the acoustic impedance is

$$\mathbf{Z} = -\frac{\omega\rho_0}{S}\frac{\mathbf{A}\ e^{-j\beta x} + \mathbf{B}\ e^{j\beta x}}{(j a - \beta)\mathbf{A}\ e^{-j\beta x} + (j a + \beta)\mathbf{B}\ e^{j\beta x}}$$

For there to be no reflected wave the horn must theoretically be infinite but in practice the reflected wave is negligible if $ka_L > 3$ where a_L is the radius of the mouth of the horn. There is already a limitation on k due to the cutoff condition and combining both requirements leads to the conclusion that

$$a_L \geqslant \frac{6}{m} \tag{5.43}$$

So, for example, if an exponential horn is to have a cutoff frequency of 100 Hz, then the flare constant, m, should be $3 \cdot 7 \ m^{-1}$ and the radius of the horn mouth should be $1 \cdot 6$ m.

If we can assume that the reflected wave is negligible then the expression for the acoustic impedance reduces to

$$\mathbf{Z} = -\frac{\omega \rho_0}{S} \frac{1}{j\alpha - \beta} \tag{5.44}$$

$$= \frac{\rho_0 c}{S} \left[j\frac{m}{2k} + \left(1 - \frac{m^2}{4k^2}\right)^{1/2} \right]$$

The impedance at the throat is obtained by putting $S = S_0$ and the real and imaginary components of $\mathbf{Z}_0$ are illustrated in Fig. 5.13.

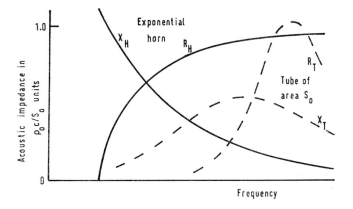

Fig. 5.13. The acoustic impedance of an exponential horn and of a tube.

Below the cutoff frequency, the impedance of the horn is entirely imaginary and waves cannot be propagated past the throat. Above the cutoff frequency, however, the impedance rapidly tends to $\rho_0 c/S_0$ and the impedance mismatch is minimal if the horn is attached to a pipe of area S_0. Therefore there is very little reflection and the waves pass through the horn. Also shown in Fig. 5.13 is the acoustic impedance of a tube of area S_0 opening directly into an infinite space and we see that this only approaches $\rho_0 c/S_0$ at much higher frequencies.

CHAPTER 6

Sound Absorption

6.1 Introduction

As a sound wave progresses through the air some absorption occurs due to viscous and molecular action. On the whole this is small except at high frequencies over large distances or in very large halls. In rooms it is the surfaces which provide sound absorption either by direct transfer of sound energy into heat energy by some internal damping process or by allowing the sound to pass through. An open window is the extreme example of this latter process and although the sound energy still exists in the absolute sense it is lost to the room and can be regarded as absorbed sound. The simplest way of defining the sound absorption of a surface is in terms of its absorption coefficient which is the ratio of absorbed energy to incident energy. It can range from a value of 1·0 for an open window to 0·001 for a very hard smooth surface such as painted plaster. The absorption coefficient is a function of frequency and of angle of incidence and at every frequency two coefficients can be quoted. One is the normal incidence absorption coefficient, α_0, which is easy to measure. The other is the random incidence absorption coefficient, $\bar{\alpha}$, which is relevant to room acoustics, where it can normally be assumed that sound waves strike the surface at all possible angles of incidence. $\bar{\alpha}$ can be measured directly or, if certain assumptions are made, it can be deduced from α_0.

Although a knowledge of the absorption coefficient is sufficient for the architect it is not sufficient for research and development work on absorbent materials. Then one needs to know the precise effect of a surface in terms of both the amplitude and phase of the reflected wave relative to the incident wave. We have already learned in Chapters 4 and 5 that in acoustics it is the ratio of pressure to particle or volume velocity which is important and that this is fully described by the impedance.

We are not concerned with changes in cross-sectional area in this chapter and it is therefore logical to describe a surface in terms of its specific acoustic impedance, that is the ratio of pressure to particle velocity at the surface. If this is known then the effect on any incident sound wave can be calculated and indeed the absorption coefficient can be deduced. Furthermore the absorbent behaviour of the surface can be understood, so allowing the designer to alter the absorption by altering the properties of the surface. Usually the surface impedance is measured but in some circumstances it can be calculated entirely from the physics of the situation.

Angle of incidence still raises some problems but if we first define the behaviour of the surface in terms of its normal impedance, that is the ratio of pressure to particle velocity for normally incident sound waves, then in most instances we can deduce the impedance for a wave incident at any other angle.

6.2 Reflection of Plane Waves at Normal Incidence

A normally incident wave $\mathbf{p}_i$ strikes the surface of normal impedance $\mathbf{z}_s$ causing it to move. As a result of the movement some of the incident wave is reflected from the surface, represented by $\mathbf{p}_r$, and some is transmitted, which we assume to be absorbed. Since we must assume that the impedance of the surface is complex it is most convenient to represent the pressures by using complex algebra. The total pressure at any point to the left of the surface is given by

$$\mathbf{p} = \mathbf{p}_i + \mathbf{p}_r$$

$$= \mathbf{A}\, e^{j(\omega t - kx)} + \mathbf{B}\, e^{j(\omega t + kx)} \qquad (6.1)$$

and the particle velocity is given by

$$\mathbf{u} = \frac{\mathbf{p}_i - \mathbf{p}_r}{z_0}$$

The ratio of pressure to particle velocity at the surface must equal the surface impedance.

therefore
$$z_s = \left(\frac{p}{u}\right)_{x=0} = \frac{A+B}{A-B}z_0 \tag{6.2}$$

At this point it is convenient to think in terms of the ratio of reflected wave to incident wave, $r = B/A$. Since B and A are both complex,

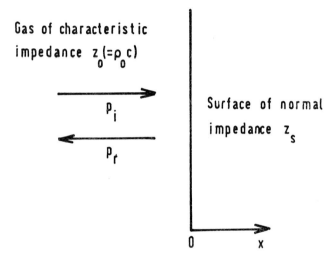

Gas of characteristic impedance $z_0(=\rho_0 c)$

p_i

p_r

Surface of normal impedance z_s

0   x

Fig. 6.1.   Normal incidence sound waves striking a surface.

r must also be complex, but it can be described fully in terms of its modulus and argument.

i.e.
$$r = B/A = |r|\, e^{j\Delta} \tag{6.3}$$

$|r|$ is the amplitude of the reflected wave relative to the incident wave and Δ is the phase shift experienced at the surface. Equation (6.2) becomes

$$z_s = \frac{1+r}{1-r}z_0 \tag{6.4}$$

which can be rearranged in the form

$$r = \frac{z_s - z_0}{z_s + {}_0z} \tag{6.5}$$

6.3   Reflection of Plane Waves at Oblique Incidence

Figure 6.2 shows a pressure wave incident, at some angle θ to the normal, on a surface of normal impedance z_s. Provided that the

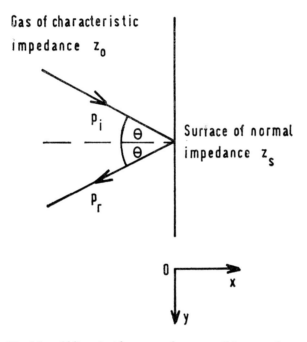

Fig. 6.2.   Oblique incidence sound waves striking a surface.

response of the surface at one point does not make itself felt at any other point, which means that surface waves are not permitted, the surface is called locally reacting and the reflected sound field can be obtained by supposing that the behaviour of one ray is identical to, but independent of, the behaviour of the other rays. Fortunately most absorbent surfaces do behave in this manner. The laws of reflection still apply so the reflected angle is equal to the incident angle and the pressure at any point to the left of the surface may be defined in terms of x and y by

$$\mathbf{p} = \mathbf{A}\, e^{j(\omega t - kx \cos\theta - ky \sin\theta)} + \mathbf{B}\, e^{j(\omega t + kx \cos\theta - ky \sin\theta)} \qquad (6.6)$$

For a perfectly smooth surface there is no need for the tangential particle velocity in the air to be equal to the tangential velocity of the surface. It is the normal component of the particle velocity, $\mathbf{u}_x$, which must be continuous, where

$$\mathbf{u}_x = \frac{\mathbf{p}_i \cos\theta - \mathbf{p}_r \cos\theta}{z_0}$$

The ratio of pressure to the x component of the particle velocity at the surface equals the normal impedance of the surface.

i.e.
$$\mathbf{z}_s = \left(\frac{\mathbf{p}}{\mathbf{u}_x}\right)_{x=0} = \frac{\mathbf{A}+\mathbf{B}}{\mathbf{A}-\mathbf{B}}\frac{z_0}{\cos\theta}$$

or
$$\frac{\mathbf{z}_s \cos\theta}{z_0} = \frac{1+\mathbf{r}}{1-\mathbf{r}} \qquad (6.7)$$

Re-arranging equation (6.7) gives

$$\mathbf{r} = \frac{\mathbf{z}_s \cos\theta - z_0}{\mathbf{z}_s \cos\theta + z_0} \qquad (6.8)$$

which is seen to be a simple extension of equation (6.5) derived for normally incident sound waves. The impedance of the surface at any angle θ is given by

$$\mathbf{z}_\theta = \mathbf{z}_s \cos\theta \qquad (6.9)$$

6.4 Measurement of Normal Impedance

We have seen in Section 6.3 that, provided a surface is both smooth and locally reacting, a knowledge of its normal impedance is sufficient to predict its effect on a plane wave impinging at any angle of incidence. Conversely, the normal impedance may be measured by subjecting the surface to a normally incident plane wave and studying the resultant sound field.

Such a measurement is most conveniently carried out in an instrument called an interferometer.

Normally the instrument is in the shape of a circular tube, although it may sometimes be rectangular. Since it is essential that only plane waves travel along the tube, there is a restriction on the maximum frequency at which any tube can be used. This is given by

$$f_{max} = \frac{c}{1 \cdot 7\, D} \qquad (6.10)$$

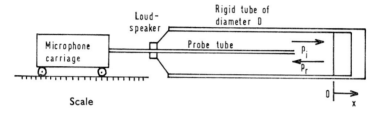

Fig. 6.3.   *Instrument for measuring impedance or absorption of a material.*

which is based on the solution for the lowest frequency cross-mode in a circular tube.

The sound field is created by a small loudspeaker and can be investigated by means of a thin probe tube attached to a microphone. The sample is usually placed just in front of a rigid backing piece, although it must be remembered that any measurements made will only apply to the particular mounting method chosen.

The sound field in the tube is given by equation (6.1) and consists of a standing wave superimposed on a travelling wave. The rms pressure along the tube as determined by the probe microphone will be of the form shown in Fig. 6.4.

The maximum values of pressure occur at points where the two waves are exactly in phase and add. Similarly the minima occur when they are exactly out of phase.

Equation (6.3) states that the modulus of the reflection ratio is given by

$$|\mathbf{r}| = \frac{|\mathbf{B}|}{|\mathbf{A}|}$$

and if the standing wave ratio in the interferometer is defined by

$$s = \frac{|\mathbf{A}|+|\mathbf{B}|}{|\mathbf{A}|-|\mathbf{B}|} = \frac{1+|\mathbf{r}|}{1-|\mathbf{r}|}$$

then
$$|\mathbf{r}| = \frac{s-1}{s+1} \tag{6.11}$$

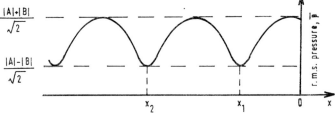

Fig. 6.4. *Pressure field inside an interferometer.*

If we re-write equation (6.1) in the form

$$\mathbf{p} = \mathbf{A}\{e^{-jkx} + |\mathbf{r}|\, e^{j\Delta}\, e^{jkx}\}\, e^{j\omega t} \tag{6.12}$$

$\mathbf{p}$ is a minimum when $-kx_m + (2m+1)\pi = \Delta + kx_m$

i.e.
$$\Delta = (2m-1)\pi - 2kx_m$$

$$= (2m-1)\pi - 4\pi x_m/\lambda$$

If x_m is taken to be the distance to the first minimum, x_1, and since $x_2 - x_1 = -\lambda/2$

then
$$\Delta = \left(1 + \frac{2x_1}{x_2 - x_1}\right)\pi \tag{6.13}$$

So by measuring the standing wave ratio and the positions of the first two pressure minima relative to the surface of the sample, equations (6.11) and (6.13) completely define the pressure reflection coefficient $\mathbf{r}$. Equation (6.4) must then be used to find the normal impedance of the surface $\mathbf{z}_s$.

This last step of going from $\mathbf{r}$ to $\mathbf{z}_s$ can be calculated or it can be done graphically by using either of the impedance charts shown in

Figs. 6.5 and 6.6. In the polar chart in Fig. 6.5, $\mathbf{r}$ is taken as a vector
with its origin at the centre. The ratio of the impedance $\mathbf{z}_s$ to z_0 is
then given by the two sets of circular coordinates, the real part
corresponding to the circles centred on the horizontal diameter of
the chart. The rectangular chart in Fig. 6.6 has two sets of circular
coordinates corresponding to $|\mathbf{r}|$ and Δ. Once plotted the real and

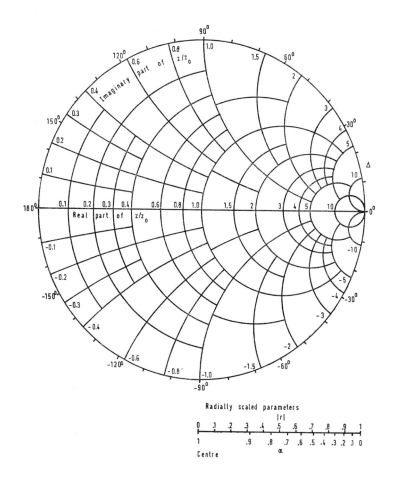

Fig. 6.5.   Polar impedance chart.

imaginary parts of z_s/z_0 can be read directly from the rectangular axes. Generally the polar chart should be used when z_s/z_0 is of the order of unity and the rectangular chart should be used when z_s/z_0 is greater than unity.

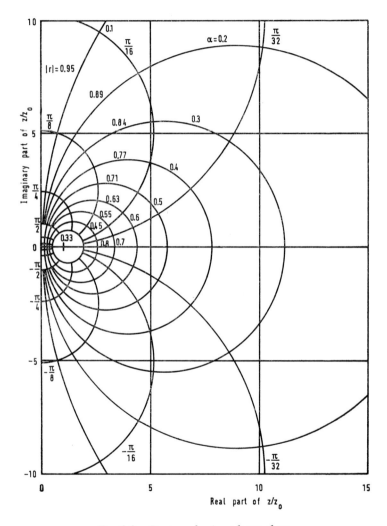

Fig. 6.6. Rectangular impedance chart.

6.5 Absorption Coefficient

The absorption coefficient of a surface is defined as the ratio of the absorbed energy to the incident energy. It is simplest to relate the energies to the incident and reflected pressure waves, in which case

$$a_\theta = 1 - \frac{|\mathbf{B}|^2}{|\mathbf{A}|^2}$$

$$= 1 - |\mathbf{r}|^2 \qquad (6.14)$$

The absorption coefficient must be a function of the angle of incidence since we have already established that the reflection coefficient is a function of the angle of incidence. The normal incidence absorption coefficient, a_0, can be obtained quite easily from the standing wave ratio measured in the interferometer in Section 6.4. If the absorption coefficient is to be obtained theoretically from a

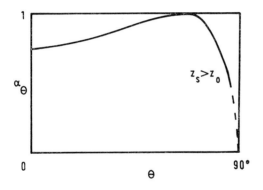

Fig. 6.7. Variation of absorption coefficient with angle of incidence.

knowledge of the surface impedance, equations (6.8) and (6.14) must be combined to give

$$a_\theta = 1 - \left| \frac{\mathbf{z}_s \cos\theta - z_0}{\mathbf{z}_s \cos\theta + z_0} \right|^2 \qquad (6.15)$$

This may be calculated or solved graphically by plotting $\mathbf{z}_s \cos\theta / z_0$ on Fig. 6.5 or Fig. 6.6.

The absorption coefficient reaches a maximum value when z_s is wholly real and $z_s \cos \theta = z_0$, but this can only occur at one value of θ and then only if $z_s > z_0$. In general the variation of α_θ with angle is of the form shown in Fig. 6.7.

Usually we are interested in the absorbing effect of a surface when it is subjected to sound waves striking it at many different angles of incidence, as, for example, when it forms one of the walls of a room. The random incidence coefficient can be found by integrating the absorbed energy with respect to θ, and dividing by the total incident energy,

i.e.
$$\alpha = \frac{\int_0^{\pi/2} I_\theta \alpha_\theta \, d\theta}{\int_0^{\pi/2} I_\theta \, d\theta} \tag{6.16}$$

I_θ is given in equation (3.42) and substitution gives

$$\bar{\alpha} = 2 \int_0^{\pi/2} \alpha_\theta \cos \theta \sin \theta \, d\theta \tag{6.17}$$

α_θ is given by equation (6.15).

Fortunately it is not necessary to work this out every time with a locally reacting material because the process is always the same and it has already been done. Fig. 6.8 shows $\bar{\alpha}$ as a function of the normal incidence coefficient α_0.

6.6   Absorption in a Room

It has been stated that many standing waves may be excited in a room and that it is easier to describe its behaviour statistically than to calculate separately the effect of each standing wave. It is assumed that the sound field is sufficiently diffuse for the pressure to be uniform throughout the enclosure except close to the walls in which case the energy density may be related to the rms pressure by the equation

$$D = \frac{\bar{p}^2}{\rho_0 c^2} \tag{6.18}$$

A perfect standing wave does not exist because whenever a sound wave strikes a boundary some of the energy is converted into heat and some can pass right through. Therefore there is always a net flow of energy towards the boundaries although it may be very small compared with the total energy in the room. Sound energy strikes any surface in the room simultaneously at all angles of incidence and

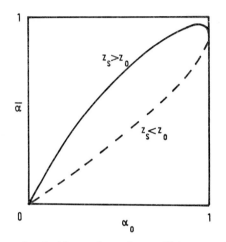

Fig. 6.8. The random incidence absorption coefficient as a function of the normal incidence absorption coefficient for a locally reacting surface.

it was shown in Section 3.7 that the total intensity on that surface can be related to the energy density in the room. The relationship is

$$I = \frac{Dc}{4} = \frac{\bar{p}^2}{4\rho_0 c} \qquad (6.19)$$

and it is assumed that I is constant throughout the room.

The rate at which energy is absorbed is equal to the intensity multiplied by the absorption within the room. Surface absorption is obtained by integrating the random incidence absorption coefficients over the total surface area of all of the boundaries. The units are either m^2 or ft^2. To this must be added the absorption of objects, such as people or chairs, which is normally quoted in terms of absorption units per object. So the total absorption is

$$A = \Sigma S_i \bar{\alpha}_i + \Sigma 0_j \qquad (6.20)$$

where S_i is the area of a particular surface of random incidence absorption coefficient $\bar{\alpha}_i$ and 0_j is the absorption of one object.

The rate of change with respect to time of the total energy in the room must be equal to the power of any source minus the rate at which energy is absorbed.

i.e.
$$\frac{d}{dt}(DV) = W(t) - IA \qquad (6.21)$$

where V is the room volume and $W(t)$ is the rate of energy emission from the source.

Substituting for I from equation (6.19) this differential equation can be written in terms of energy density

i.e.
$$\frac{dD}{dt} + \frac{Ac}{4V}D = \frac{W(t)}{V} \qquad (6.22)$$

Without defining $W(t)$ more precisely the solution to this equation can only be written in the form of an integral

i.e.
$$D = \frac{1}{V}e^{-(Ac/4V)t} \int_{-\infty}^{t} e^{(Ac/4V)\tau}W(\tau)\,d\tau \qquad (6.23)$$

Because of the exponential within the integral the energy density at any instant t only depends on the behaviour of $W(\tau)$ in the $4V/Ac$ seconds immediately prior to that instant.

If $W(\tau)$ is constant over this period the solution becomes

$$D = \frac{4W}{Ac} \quad \text{and} \quad I = \frac{W}{A} \qquad (6.24)$$

and we see that the energy density is constant within the room and is inversely proportional to the total absorption.

If the sound source suddenly starts at time $t = 0$ the solution becomes

$$D = \frac{1}{V}e^{-(Ac/4V)t} \int_{0}^{t} e^{(Ac/4V)\tau}W\,d\tau$$

$$= \frac{4W}{Ac}[1 - e^{-(Ac/4V)t}] \qquad (6.25)$$

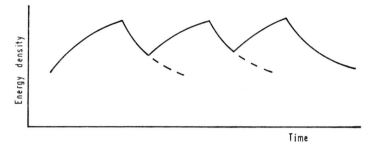

Fig. 6.9. Energy density in a room resulting from a series of pulses.

and if the sound source suddenly stops at time $t = 0$ the solution becomes

$$D = D_0\, e^{-(Ac/4V)t} \qquad (6.26)$$

The response of the room to a succession of short pulses, for example, the individual syllables when someone is speaking, is given by equations (6.25) and (6.26) and is illustrated in Fig. 6.9.

The effect of the room is to slur the pulses or syllables together, which is called loss of articulation, and if the pulses are produced at a fixed rate the amount of slurring depends upon the ratio A/V appearing in the exponentials. The smaller the total absorption and the larger the room the more the pulses tend to run together.

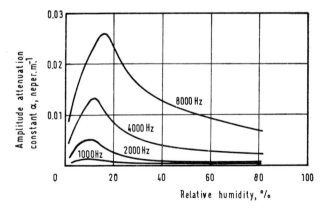

Fig. 6.10. Air absorption at 20°C. Energy attenuation constant $m = 2\alpha$.

So far it has been assumed that all of the absorption takes place at the boundaries. This is not always true, because at frequencies above 4000 Hz the air within the room absorbs a significant proportion of the sound energy. The air absorption is a function of humidity and frequency and the amplitude attenuation constant, a, is shown in Fig. 6.10.[4] Energy is proportional to the square of the amplitude and so the energy attenuation constant m is equal to $2a$.

Without any other absorption sound energy is dissipated according to the equation $D = D_0 e^{-mx}$, or substituting $x = ct$, $D = D_0 e^{-mct}$. The rate of change of energy may be found by differentiating and the term $-mcDV$ should appear on the right hand side of equation (6.21). Equation (6.22) becomes

$$\frac{dD}{dt} + \frac{(A+4mV)c}{4V} D = W(t)$$

and the total absorption increases from A to $(A+4mV)$ throughout equations (6.23) to (6.26).

6.7   Reverberation Time

The fact that sound energy in a room decays exponentially when the sound source is stopped is called reverberation. If the logarithm of the pressure or the energy is plotted as a function of time the exponential becomes a straight line.

In order to specify the reverberation quantitatively, the reverberation time is defined as the time taken for the energy to decay to one millionth of its initial value. On the logarithmic scale this corresponds to a decrease of 60 dB either in terms of the energy level or the sound pressure level.

Ignoring the air absorption this definition may be applied to equation (6.26) to give

$$D = 10^{-6}D_0 = D_0 e^{-(Ac/4V)T}$$

where T is the reverberation time. Taking logarithms gives

$$\frac{cAT}{4V} = 2{\cdot}3 \times 6$$

i.e.
$$T = \frac{55}{c} \cdot \frac{V}{A} \qquad (6.27)$$

Equation (6.27) is non-dimensional but it is usual to substitute values for the velocity of sound, c, in either ft. sec^{-1} or msec^{-1}. This gives the two alternative forms of the Sabine equation.

$$T = \frac{0 \cdot 049 \ V}{A} \quad \text{where} \quad V \text{ is in ft}^3$$
$$\text{and} \quad A \text{ is in ft}^2$$

or

$$T = \frac{0 \cdot 16 \ V}{A} \quad \text{where} \quad V \text{ is in m}^3$$
$$\text{and} \quad A \text{ is in m}^2$$

(6.28)

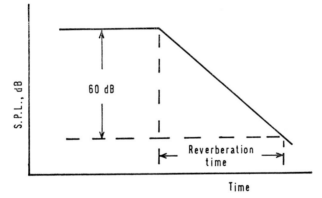

Fig. 6.11. *Decay of sound pressure level in a reverberant room.*

If air absorption is significant equation (6.27) becomes

$$T = \frac{55}{c} \cdot \frac{V}{A + 4mV} \qquad (6.29)$$

and equations (6.28) are modified accordingly.

The equations which have been developed only apply to the limiting case of the total surface absorption in the room being small. If the mean absorption coefficient of all the surfaces, given by

$$\bar{\alpha}_m = A/S \qquad (6.30)$$

where S is the total surface area, is greater than about 0·1, it is more accurate to use the Norris-Eyring equation. At every reflection the intensity is reduced by the factor $\bar{\alpha}_m$ and so after n reflections the remaining intensity is given by

$$I = I_0(1 - \bar{\alpha}_m)^n$$

The mean free path is $4V/S$ and so the number of reflections in time t is $Sct/4V$.

therefore
$$I = I_0(1 - \bar{\alpha}_m)^{Sct/4V}$$

Using the definition that the reverberation time T is the time taken for the intensity to decrease to $10^{-6}I_0$

$$(1 - \bar{\alpha}_m)^{(Sc/4V)T} = 10^{-6}$$

therefore
$$\frac{Sc}{4V} Tln(1 - \bar{\alpha}_m) = -6 \times 2 \cdot 3$$

therefore
$$T = \frac{55}{c} \cdot \frac{V}{-Sln(1 - \bar{\alpha}_m)} \tag{6.31}$$

Air absorption is unlikely to be very important in this case, but, if it is necessary, it can be included by adding the term $4mV$ into the denominator as before.

Now that reverberation time has been defined, and it is obviously possible both to measure and calculate it, it is worth considering its subjective effect. It was shown in the previous section that a long reverberation time would result in loss of articulation. On the other hand reverberation will tend to amplify the sound level and so make the speaker more easily heard. Consequently the most satisfactory reverberation time varies slightly with the room volume.

Before the nineteenth century music tended to be written for performance in existing rooms. For example, organ music was written for churches which, being large and lacking soft furnishings, had very long reverberation times. Chamber music, on the other hand, was written for performances in well furnished private houses and emphasis could be placed on clarity. Wagner reversed this trend by having the Opera House at Bayreuth built specifically for the performance of his music. Nowadays, a new concert hall is built to suit the style of music for which it is intended, and those people involved or interested in music have very specific ideas on the requirements.

It should not be thought that the reverberation time is the only criterion in room acoustics but it is probably the most important and a list of customary or recommended times is given in Table 6.1. The times for concert halls and lecture halls are shown again in Fig. 6.12 as a function of the room volume.[1]

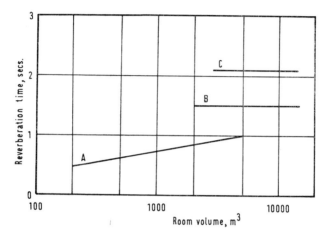

Fig. 6.12. Optimum reverberation time at 500 Hz as a function of room volume. A for a lecture hall; B for classical and modern music; C for romantic music.

TABLE 6.1

TYPICAL REVERBERATION TIMES AT 500 Hz

Concert hall .	.	.	1·5–2·1 seconds
Opera house .	.	.	1·1–1·3 seconds
Lecture hall .	.	.	0·5–1·0 seconds
Recording studio	.	.	0·4–0·5 seconds
Living room .	.	.	0·5 seconds

The reverberation time should be reasonably constant over the frequency range 63 to 2000 Hz. At higher frequencies a reduction is acceptable.

6.8   Random Incidence Absorption Measurement

The random incidence absorption coefficient can be deduced from normal incidence measurements if the material can be classified as 'locally reacting'. Nevertheless it would be better if the random incidence absorption coefficient could be measured directly. The interferometer is suitable for comparative measurements because of its simplicity and cheapness but the final test must be made under random incidence conditions. This can most easily be done in a specially constructed reverberant chamber in which the total absorption can be related to the reverberation time by means of the Sabine equation. The first measurement is made with the room empty which gives

$$T_1 = \frac{0 \cdot 16 V}{S_t \bar{\alpha}_1} \qquad (6.32)$$

where S_t is the total surface area of the walls, floor and ceiling and $\bar{\alpha}_1$ is the absorption coefficient of these surfaces. Provided that the room is initially very reverberant, the introduction of a small area S_A of material whose absorption coefficient is $\bar{\alpha}_2$ will cause a marked decrease in reverberation time given by

$$T_2 = \frac{0 \cdot 16 V}{(S_t - S_A)\bar{\alpha}_1 + S_A \bar{\alpha}_2} \qquad (6.33)$$

Equations (6.32) and (6.33) can be combined to give

$$\bar{\alpha}_2 - \bar{\alpha}_1 = \frac{0 \cdot 16 V}{S_A} \left(\frac{1}{T_2} - \frac{1}{T_1} \right) \qquad (6.34)$$

and usually $\bar{\alpha}_1$ is sufficiently small when compared with $\bar{\alpha}_2$ to be ignored. If not it may be obtained very easily from equation (6.32).

The frequency range of interest is from 125 to 4000 Hz, and the optimum size of the room is about 200 m³. If it were any smaller the sound field would be unsatisfactory at 125 Hz and if it were much larger the air absorption could become appreciable at 4000 Hz. Ideally no two surfaces should be parallel in order to provide good diffusion and sometimes non-absorbent reflectors are hung in the room. The sound field is produced by a loudspeaker system which should be placed in one corner of the room in order to excite the maximum possible number of room modes. Even then a single frequency is not satisfactory because the result would be too dependent on the room shape. It is necessary to use either a warble tone deviating from the centre frequency by $\pm 10\%$ at a rate of 5 to 10 Hz, or a $\frac{1}{3}$ octave band of white noise. For example, if the centre frequency were to be 1000 Hz the warble tone would excite all those room modes with natural frequencies in the range 900 to 1100 Hz, while the $\frac{1}{3}$ octave band would excite all those in the range 890 to 1122 Hz. The sound field is still not completely diffuse at the lowest frequencies and it is common practice to measure the reverberation time at several different positions at each frequency. The measurements are usually made at octave intervals starting at 125 Hz.

6.9 Common Absorption Treatments

The intention here is not to discuss the virtues of particular commercial absorbents but rather the principles underlying different basic types of absorber. They may be split into six different categories as illustrated in Fig. 6.13.

(a) *Rigid Porous Layer*

The sound waves are able to penetrate the material directly and it is the frictional forces experienced by the air molecules which absorb the energy. In this particular case the sound absorption may be related to the direct flow resistance. There are unfortunately two conflicting requirements. Before the energy can be absorbed the sound wave must be able to enter the porous layer and for maximum transmission the impedance (and therefore the flow resistance) should be low. There is then a finite distance for the wave to travel before it strikes the solid wall and is reflected. For maximum absorption in this finite distance the impedance (and flow resistance) should be high, (*see* Fig. 6.14). Consequently for maximum absorption with a given thickness of material there is an optimum flow resistance.

Absorption is poor at low frequencies because maximum energy dissipation occurs in the region of maximum particle velocity and this is farther from the wall at low frequencies. The obvious way to

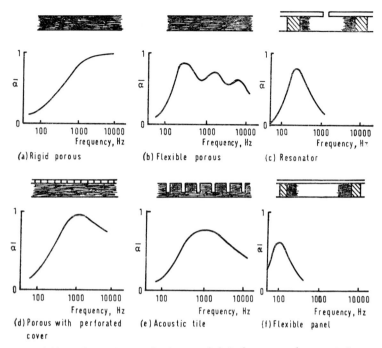

Fig. 6.13.   *Absorption mechanisms and their frequency characteristics.*

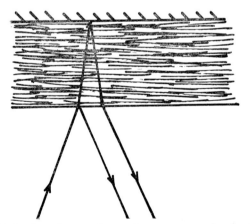

Fig. 6.14.   *Reflection of a sound wave from a porous absorbent layer on a rigid backing.*

remedy the situation is to use a thicker layer of absorbent material but that tends to be uneconomical. A more acceptable method is to leave an air gap between the absorbent and the wall which increases absorption at low frequencies with some reduction of absorption at higher frequencies.

Great care must be taken with the surface of this type of absorbent. Apart from the danger of surface damage there is the distinct possibility that dirt and dust particles can clog the pores, so impeding

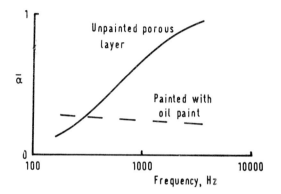

Fig. 6.15.   *Absorption coefficient of a rigid porous layer.*

the airflow and reducing the absorption. A coat of paint could be absolutely fatal (*see* Fig. 6.15) unless care is taken to use a special paint which will not form a continuous layer but will only colour the edges of the pores.

(b) *Flexible Porous Layer*

This type of material may only be pseudo-porous, *i.e.* although it may be 70 or 80 % air, the air may be enclosed in many tiny cells and the direct flow resistance is infinite. Taking this extreme case first, it is evident that any absorption which occurs must be due to vibrations of the absorbent structure and it is the elastic properties which become important. It is a multi-degree of freedom system, and the impedance becomes a minimum at the frequencies of normal modes of vibration. The energy will be absorbed by the internal damping of the system and it is that which determines the width of the peaks in the absorption curve shown in Fig. 6.13 (b).

An air gap behind the absorbent will modify the absorption curve in so far as the boundary conditions of the vibrating material are modified. Since this amounts to a reduction of stiffness it might be expected that the absorption peaks would shift to lower frequencies.

Materials which are both flexible and truly porous provide viscous and vibrational absorption. At high frequencies the viscous absorption predominates while at low frequencies flexibility may supply absorption peaks (*see* Fig. 6.16.).

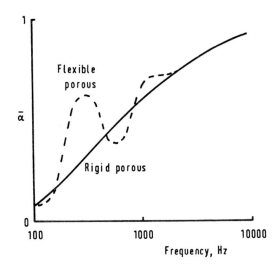

Fig. 6.16.   *The effect of flexibility on the absorption of a porous layer.*

An air space behind the layer will again be beneficial at low frequencies, but dirt or paint can have a very detrimental effect at higher frequencies.

(c) *Single Resonator*

The simplest form is the Helmholtz resonator which is an enclosed volume of air connected to the room by a small opening or neck. All of the dimensions are small compared with the wavelength of sound at the frequency at which the system is designed to operate. Its action has already been analysed in Section 5.2 where it was shown that the system resonates at a frequency given by

$$f_0 = \frac{c}{2\pi} \left(\frac{\pi a^2}{V(l + 16a/3\pi)} \right)^{1/2} \tag{6.35}$$

It is quite easy to design a resonator with a low resonant frequency and at that frequency the impedance presented to any sound pressures acting on the mouth of the resonator reaches a minimum value equal to the damping in the system which has so far been ignored. Consequently it will take in energy from any impinging sound wave of the appropriate frequency, and as would be expected, maximum absorption will occur if the impedance is equal to $\rho_0 c$. Care must be taken that the damping is not less than this for two reasons.

(1) The decay rate of the resonator could be less than the decay rate of the room and consequently, when the source is stopped, the resonator would become a decaying sound source.

(2) The resonator could be too sharply tuned and absorption would only take place over a frequency range too narrow to be of any use.

It is, therefore, common practice to increase the damping artificially by introducing some loose fibrous material, such as glass fibre or mineral wool, into the system. Fig. 6.17 gives some indication of how the absorption depends upon the damping.

If too much fibrous material is introduced the system ceases to be resonant and absorption occurs by the processes described under headings (a), (b) or (d).

The slit resonator is a complex form of the Helmholtz resonator in which one dimension is comparable with the wavelength of sound at the frequencies involved. Architecturally it is much more attractive than the smaller resonator both in appearance and construction and its mode of operation is very similar.

(d) *Porous Layer with Perforated Covering*

The perforated covering of 30% or more open area such as is used in silencers or on walls simply to keep the porous layer in place does not affect the absorption of the porous layer at all. This section is concerned with coverings of 15% or less perforation and such a system may be treated theoretically as a case of multiple Helmholtz resonators. The porous layer is necessary because of the reasons given in Section (c), and also to prevent the individual resonators from coupling together. This would not matter with normal incidence sound waves but could be undesirable with oblique incidence waves

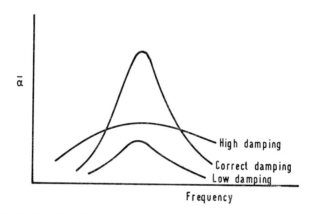

Fig. 6.17. The effect of damping on the absorption of a resonator.

because of phase differences. The design of perforated panel absorbents is well documented, but too lengthy to discuss in any detail in this book.

(e) *Acoustic Tile*

This usually takes the form of a compressed rigid fibre panel, the surface of which is perforated with a great number of holes, slots or cracks. The original specification called for holes of 4·3 mm diameter, with about 15 mm between holes and this formula is still in use. There seems to be no real theory for this type of material but it has been suggested that absorption is mainly due to the resonance of the air within the main holes. If the material were otherwise non-porous this would occur at a frequency such that the depth of the hole corresponded to a quarter-wavelength. The porosity of the material,

however, considerably reduces the effective stiffness of the holes and
so reduces the resonant frequency. The peak is fairly broad because
of the large amount of damping which is present.

The surface area around the holes may be treated or painted in
any way, which gives this type of material a great advantage and it is
in widespread use especially on ceilings.

(f) Flexible Panel

An impervious panel mounted a short distance from a rigid wall
has a resonant frequency dependent upon the mass of the panel and
the stiffness of the air in the cavity, given by

$$f = \frac{1}{2\pi} \left(\frac{\gamma P_0}{Md}\right)^{1/2} \tag{6.36}$$

where M is the mass per unit area of the panel and d is the thickness
of the cavity.

It is found that the flexural stiffness of the panel itself is negligible.
At the resonant frequency energy will be absorbed by internal losses
within the panel itself and by friction around the edges. As with the
single resonator this damping may be insufficient in which case the
gap should be filled with some cheap fibrous substance.

This type of absorber is only suitable for use at low frequencies
and even then it is difficult to use in practice. The main source of
trouble is the seal which is necessary to maintain the correct air
stiffness.

CHAPTER 7

Hearing, Loudness and Criteria

7.1 The Human Ear

Most of the science of audio-acoustics must at some time be related to human subjective impressions. It is therefore relevant to consider briefly the working of the ear and then to relate subjective impressions to objective measurements.

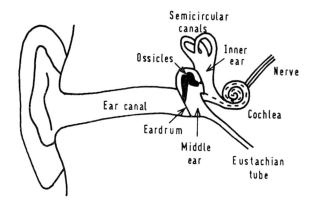

Fig. 7.1. Schematic cross-section of the ear.

Fig. 7.1 shows a cross-section of the ear, which consists of three main parts. The most important part is the inner ear, the other two parts can be considered merely as accessories to this. The parts will be described from the exterior inwards.

The outer ear serves to concentrate the pressures onto the drum

130

and assists in ascertaining the direction from which the sound is coming.

The middle ear contains air and is connected through the Eustachian tube to the external atmosphere. This is normally closed but can be opened by yawning or swallowing so that slow changes in atmospheric pressure, as in an aeroplane ascending or descending, can be communicated to the middle ear to relieve the pressure differential across the eardrum. Cyclic pressure changes associated with sound waves are not communicated through the Eustachian tube, and consequently the eardrum moves with the fluctuating pressure in the outer ear. The middle ear also contains some small bones which provide a mechanical connecting link between the eardrum and the tympanic membrane. This link serves to reduce the amplitude of motion but to increase the force.

The inner ear, beyond the tympanic membrane, is filled with liquid and it is here that the nerves respond to the pressure fluctuations to create the sensation of hearing. A membrane divides the inner ear into two for almost its entire length and different portions respond to different frequencies of sound. Hence the ear is able to discern both frequency and pressure. Masking of one sound by another louder sound often occurs and it is due to interaction at the membrane in the inner ear. In general the higher frequencies are masked by the lower frequencies and the degree of masking depends upon the frequency difference and upon the relative magnitudes of the two sounds.

The whole process of hearing is evidently very complicated and so it is unreasonable to expect a simple relationship between a subjective impression and the response of an objective measuring system in any given sound field.

7.2 Loudness of Pure Tones

Numerous experiments have been carried out in which many people have been asked to assess the loudness of pure tones of known frequency and pressure. It is essential that the subjects should have nothing physiologically wrong with their hearing and that there should be no history of exposure to high noise levels; in other words the subjects' hearing should be normal. A test may be carried out in two different ways; either the subject wears earphones in which case the actual pressure level on the ear is measured or else the subject sits in an anechoic room with the sound source directly in front of him and the free field sound level to which he is exposed is measured separately when the subject is absent. In either case the normal procedure is to use a 1000 Hz tone as a reference and to ask the subject to adjust the

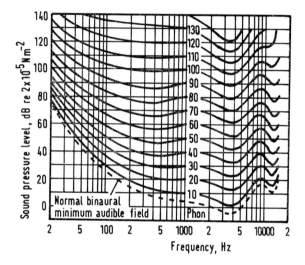

Fig. 7.2. Equal loudness contours for pure tones. Free field conditions with frontal sound.

volume of a tone of different frequency until it is of equal loudness to the reference signal. The frequency is then altered until the entire audio frequency range has been covered. The individual points are plotted on a graph of sound pressure level, derived either from the pressure in the earphones or from the free field pressure, versus frequency and joined up to give an equal loudness contour. This is repeated for different levels of the reference tone. The exception to this procedure is the determination of the threshold contour which is measured absolutely at all frequencies.

The free field method is probably more representative of real conditions and the corresponding contours, illustrated in Fig. 7.2,[5] are generally used as the standard for estimating the loudness of different sounds. The pressure method is, however, much simpler to carry out and the corresponding pressure contours are used as the standard when a person's hearing is being checked by means of earphones in an audiometric booth.

The loudness level is measured in phons and for convenience the numerical value of loudness level is equal to the numerical value of the sound pressure level at 1000 Hz. The threshold of hearing for the average person corresponds to zero phons when wearing earphones but is a little higher in free field conditions. The 120 phon contour is the maximum normally experienced since the sensation becomes painful at such high levels. In general we see that the ear is most sensitive in the frequency region between 500 and 10,000 Hz. As the frequency decreases below 500 Hz the sensitivity steadily decreases until at 20 Hz and below the sound is not heard as a pure tone but rather as a succession of pressure variations. At the other end of the frequency range the average ear does not respond at all above about 16,000 Hz and for older people this may drop to 12,000 Hz.

The loudness levels are useful for comparing two different frequencies of equal loudness but the numerical values are still not directly related to subjective impressions. For example, 60 phons is not twice as loud as 30 phons. In fact over most of the audible range a doubling of loudness corresponds to an increase of 10 phons. For this reason the concept of loudness index (measured in sones) has been introduced and the relationship between phons and sones is shown in Fig. 7.3.[6] There is now a direct relationship between sones and loudness, *i.e.* 8 sones is twice as loud as 4 sones, four times as loud as 2 sones and eight times as loud as 1 sone. Nevertheless it is still customary to describe the loudness of a sound not in sones but in phons.

7.3   Loudness of Noise

In most instances it is necessary to assess the loudness of wide band noise rather than of pure tones. One way of doing this is to split the noise into different frequency bands, *e.g.* octaves, and measure the sound pressure level of each band in decibels. Using our knowledge of the sensitivity of the ear, the measured level in each band can be converted to a subjective loudness level in phons (or sones). Finally the loudness levels of the individual bands must be recombined so that the total subjective effect is achieved. Because of masking this is not just a straightforward arithmetic addition but rather a process of addition wherein the louder bands are given greater weight than the others.

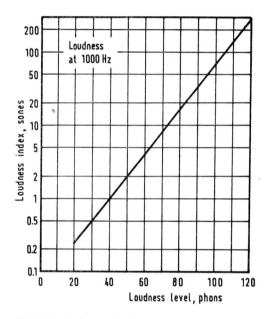

Fig. 7.3.   Relationship between phons and sones.

S. S. Stevens[7] has developed an empirical method for doing this which starts with a set of equal loudness contours, similar to those in Fig. 7.2, but obtained from tests using octave bands of diffuse sound, that is sound containing many different frequencies within a given range and coming from all directions at the same time. These are shown in Fig. 7.4. The contours have been labelled by loudness index values in sones although phons could have been used. The relationship between phons and sones is the same as in the previous section. So each octave band of the noise to be assessed is ascribed a loudness index according to Fig. 7.4 and the summation is carried out according to the following rule

$$S_t = S_m + 0\cdot3(\Sigma S - S_m) \tag{7.1}$$

where S_t is the total loudness in sones, S_m is the loudness in sones of

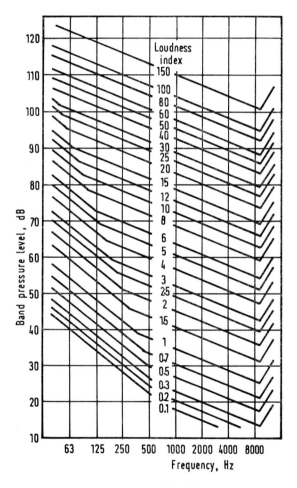

Fig. 7.4. Stevens' equal loudness contours.

the loudest octave band and ΣS is the sum of the loudness in sones of all of the octave bands. S_t is then converted back to phons using Fig. 7.3 to give the result in Stevens' phons (OD). (OD) indicates that octave bands of diffuse sound have been used and implies the use of Stevens' method. The constant 0·3 takes into account the band width and the masking of the remaining bands by the loudest band.

An alternative method has been proposed by the German scientist, E. Zwicker.[8] This is based on a theoretical model of the human ear and again follows the general principles of Stevens' method. The total noise is split into different frequency bands (or groups) and each is ascribed a certain loudness. The bands are narrower than in Stevens' method, $\frac{1}{2}$ octaves being used at low frequencies and $\frac{1}{3}$ octaves elsewhere. Masking of each band is allowed for separately according to its loudness in relation to the rest and then they are summed. Usually a special printed sheet is used which enables the loudness, masking and summation all to be performed graphically. There is a choice of pre-printed sheets according to whether the sound is frontal or diffuse and the total loudness in sones is converted to either phons (GF) or phons (GD). (GF) refers to groups of frontal sound and (GD) refers to groups of diffuse sound. Either implies the use of Zwicker's method.

7.4 Noisiness

Noisiness and loudness are not the same, the difference being that loudness is an absolute physical thing and noisiness is a combination of loudness and circumstance. It would still seem reasonable to use the loudness units described in the previous section, although it must be remembered that a sound of a certain loudness in one place may appear noisy while in a different place the same sound may not be noticed. So in conjunction with the units of loudness there should also be a set of criteria which take into account the circumstances in which the sound is heard.

There is, however, a school of thought which claims that the problem cannot be treated so simply. Instead, each type of noise source and set of circumstances should be considered separately and a method developed for measuring the noisiness directly. Generally this would be impracticable but it has been developed by K. D. Kryter[9] in the single case of aircraft noise affecting residential areas. Equal noisiness contours were established by asking groups of

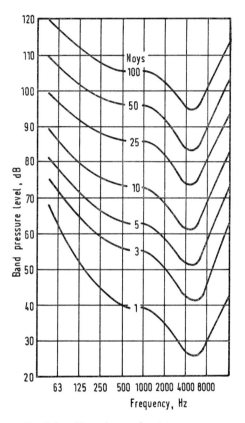

Fig. 7.5.   *Kryter's equal noisiness contours.*

subjects how noisy they found various aircraft noises, and these are shown in Fig. 7.5. The noise index is measured in noys, which are similar in concept to sones, *i.e.* a noise of 4 noys is twice as noisy as one of 2 noys.

As with Stevens' method of assessing loudness, the noise is

analysed into octave bands, each band is given a noise index according to Fig. 7.5 and the bands are added according to Stevens' rule.

i.e.
$$N_t = N_m + 0.3(\Sigma N - N_m) \qquad (7.2)$$

where N_t is the total noise index in noys, N_m is the noise index of the noisiest band in noys and ΣN is the sum of the noisiness in noys of all of the octave bands. Finally the noys are converted to PN dB (perceived noise level in decibels) which are analogous to phons. The relationship between PN dB and noys is exactly the same as that between phons and sones, so Fig. 7.3 is used for the conversion.

7.5 The Use of Weighting Networks

The methods just described for calculating loudness or noisiness require either a lengthy analysis followed by a certain amount of

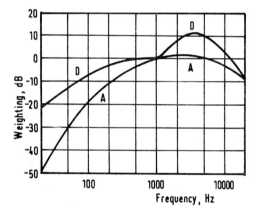

Fig. 7.6. Weighting curves.

computational work or a very expensive instrument which will both analyse and compute. A much simpler method which would give an immediate result without using expensive equipment would be very attractive. The main problem is to account for the variation in sensitivity of the ear with both frequency and level, and to this end three electrical weighting networks have been agreed upon. These are referred to as the A, B and C weighting networks and their shapes correspond roughly to the 40, 70 and 100 phon equal loudness contours of Fig. 7.2. The idea was that one of the networks should be switched in between a microphone and a meter according to the noise level being measured and this would reduce the effect of low and high frequencies relative to mid-frequencies in the same way as the ear does. In fact the use of three weighting networks was found to be unsatisfactory but if the A weighting network is used irrespective of the noise level the results obtained are quite meaningful for most types of noise. A possible exception is aircraft noise and it has been suggested that a special D weighting curve should be used in this particular case. Its shape corresponds approximately to the 50 noys contour in Fig. 7.5. The A and D weighting curves are illustrated in Fig. 7.6 and when these are used for measurements the results are given in dB(A) and dB(D) respectively. A noise level measured in dB(D) is numerically 10 less than the same noise level in PN dB.

7.6 Noise Rating Curves

Although simple, the use of weighting networks gives no indication whatever of the frequency content of the noise. This can be a great disadvantage if it is necessary to reduce the noise, because the engineer will not know which particular frequencies must be reduced most. A method of noise assessment which is popular in both the U.S.A. and in Europe is to analyse the noise into octave bands and then to plot these octave band levels on special graph paper on which noise contours are already printed. Those in use in the U.S.A. are referred to as noise criteria but the Europeans have their own set of noise rating curves which are shown in Fig. 7.7.[10]

Once the octave band levels have been plotted on the graph paper each octave band level can be assigned a noise rating number (N) and the noise rating of the noise as a whole is equal to the highest individual octave band noise rating. Consequently the noise rating number of a noise cannot be expected to correlate as well with subjective reaction as, say, the loudness in Stevens' phons, but the usefulness of the method in noise control problems more than compensates for this deficiency.

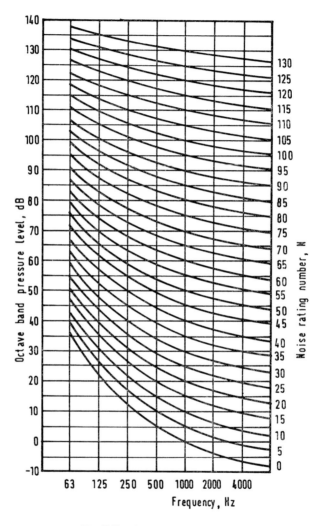

Fig. 7.7. Noise rating contours.

7.7 The Choice of Method

The various methods for assessing loudness or noisiness may be summarised in order of increasing complexity as follows:

(1) The use of a weighting network which gives a single figure reading in dB(A) or dB(D). No information is obtained about the frequency content of the noise.

(2) The use of noise rating curves, which requires analysis of the noise into octave bands. The noise rating number is then assumed to be equal to the noisiest octave band and all other octave bands are ignored.

(3) The calculation of Zwicker's Phons, Stevens' Phons or PN dB which also requires analysis of the noise into frequency bands. The individual bands are summed according to a special rule which gives most weight to the loudest or noisiest band but also includes a contribution from the other octave bands.

The British recommended method for measuring noise, other than aircraft noise, is to use the A weighting network. This has obvious advantages in many situations, *e.g.* for police officers who need to assess the noisiness of motor vehicles or public health inspectors who need to assess industrial noise, because there is neither the time nor the facility for more detailed analysis. The engineer, however, who may be called upon to reduce the noise, must be able to analyse it into octaves and, on occasions, into narrower bands. Once analysed, the noise rating method gives a very clear indication of the reduction required in each octave band in order that a specified noise rating should not be exceeded. Zwicker's method is marginally more accurate than Stevens' method for assessing loudness, but in most cases there would be no practical advantage in using these units for estimates of noisiness over the much simpler A weighting method. The recommended procedure for assessing aircraft noise is to measure the level in PN dB, but, because of the complexity, the simpler D weighting network is already being used in some instances. The loss of accuracy is small and it might even be argued that the D weighting network is only marginally superior to the A weighting network even in this sphere. It may be concluded therefore that, however empirical and academically unsupportable it may be, the A weighting network gives almost as reliable an estimate of subjective reaction to noise as any method which has been developed, and the prospect of using one simple measurement unit for all types of noise has undeniable appeal.

It is occasionally necessary to transfer from one system of units to another, *e.g.* if the noise level is available in one form and the criterion to be met is only available in another. There is no absolute method of transference, but the following approximate numerical equivalents may be assumed.

$$\text{Noise rating number} \equiv \text{dB(A) level} - 5 \qquad (7.3)$$

$$\text{Zwicker Phon level} \equiv \text{Stevens' Phon level} \equiv 1.05 \times \text{dB(A) level} + 10 \qquad (7.4)$$

$$\text{PN dB level} \equiv \text{dB(D) level} + 10 \equiv \text{dB(A) level} + 14 \qquad (7.5)$$

7.8   Criteria for Damage to Hearing

Constant exposure to high noise levels can produce permanent damage to a person's hearing. This should not be confused with a temporary threshold shift occurring after a short exposure to a moderately high noise, which is immediately discernible to the individual but from which he recovers after two or three hours. A permanent threshold shift is far more insidious since it occurs very gradually over perhaps two or three years and is not obvious to the individual until it has reached an advanced stage. The damage is to the nerves in the inner ear and results in a loss of sensitivity at high frequencies, most markedly at about 4000 Hz. This means that sibilants are not clearly heard, making the understanding of conversation very difficult. Hearing tests made every few months would show whether or not any permanent damage was occurring but at the same time it should be possible to assess in advance whether or not a particular noise climate is likely to cause damage. A considerable amount of research work has been done and Table 7.1 shows for how long a person can be exposed to various continuous noise levels without great risk of damage. It should be remembered that these exposures are based on an average reaction and it would still be wise to carry out regular hearing tests in borderline cases.

TABLE 7.1
SUGGESTED MAXIMUM EXPOSURE TIMES FOR AVOIDING
DAMAGE TO HEARING

Noise level in dB(A)	Noise rating number	Permissible exposure time in min day^{-1}
90	85	500
95	90	140
100	95	50
105	100	30
110	105	17
115	110	10

7.9   Criteria for Annoyance

(a) *Buildings Other Than Dwellings*

From the point of view of comfort and convenience there is a limit to the tolerable background noise level which depends upon the situation. There are, as yet, no British recommendations but there

TABLE 7.2
RECOMMENDED MAXIMUM NOISE LEVELS

Type of room	American criteria in dB(A)	European criteria	
		N	dB(A)
Broadcasting studio . . .	25–30	15	20
Concert hall.	25–30	20	25
Legitimate theatre (500 seats) . .	30–35	20	25
Classroom, music room, TV studio .	35	25	30
Courtroom	30–35	30	35
Library	40–45	30	35
Cinema, hospital	40	30	35
Church	35	30	35
Restaurant	55	45	50
Private office	40–45	40	45
General office	60–65	55	60
Workshop	—	65	70

are American noise criteria recommendations which have been converted to dB(A) for inclusion in Table 7.2 and there are also European noise rating recommendations which are given in terms of both noise rating numbers and dB(A) levels in Table 7.2.

(b) Dwellings

There is a specific British Standard for rating industrial noise affecting mixed residential and industrial areas.[11] It is based on experience of what people will tolerate in varying circumstances and is particularly helpful in anticipating reaction to a new noise although it can be used for checking whether or not a complaint about existing noise is reasonable. The noise level in dB(A) outside a particular residence is measured or estimated and is compared with either the background noise level, which exists apart from the noise in question, or a criterion based on the locality, time and season. Corrections are made for the type and duration of the noise.

Aircraft noise has received some attention in the vicinity of airports and a special unit has been evolved,[12] known as the noise and number index (NNI). It is based on the number of aircraft in a twelve-hour period (N) and on the average peak noise level measured in PN dB.

i.e. $\quad$ NNI = (average peak noise level) $+ 15 \log_{10} N - 80$ $\quad$ (7.6)

An external NNI of 55 has been taken as the recommended limit in the vicinity of London Airport and householders exposed to more noise have been offered a special grant for soundproofing their homes. This practice has not yet been adopted elsewhere in Britain.

Traffic noise has recently been attracting some interest. The unit of measurement is dB(A) but account must also be taken of the fluctuation of noise level with time. A simple way of doing that is to measure the levels in dB(A) which are exceeded for 10 % and 90 % of the time. If these are designated L_{10} and L_{90} the traffic noise index (TNI) is defined as follows.[13]

$$TNI = L_{90} + 4(L_{10} - L_{90}) - 30 \qquad (7.7)$$

It has been suggested that an external TNI of 74 should be adopted as a criterion but further research is still required.

Appendix

Substance	Density $kg\ m^{-3}$ ρ_0	Velocity $m\ sec^{-1}$ c	Characteristic impedance $Nm^{-3}\ sec$ $\rho_0 c$		
Air . . .	1·21	343	415		
Carbon dioxide .	1·84	267	481		
Hydrogen . .	0·084	1330	112		
Nitrogen . .	1·17	349	409		
Oxygen . . .	1·33	326	434		
			$\times 10^6$		
Pure water . .	998	1483	1·48		
Sea water . .	1025	1522	1·56		
		Bulk	Rod	Bulk	Rod
				$\times 10^6$	$\times 10^6$
Aluminium . .	2700	6370	5100	17·2	13·8
Brass . . .	8400	4370	3450	36·7	29·0
Crown glass . .	2500	5660	5340	14·1	13·4
Perspex . .	1190	2700	2180	3·21	2·60
Hard rubber . .	1100	2400	1450	2·64	1·59
Mild steel . .	7800	5960	5200	46·4	40·5

145

TABLE A.2
CENTRE AND LIMITING FREQUENCIES OF
1/1 AND 1/3 OCTAVE BANDS[14]

1/1 Octave band centre frequencies	Limiting frequencies	1/3 Octave band centre frequencies
	22	
		25
	28	
31·5		31·5
	36	
		40
	45	
		50
	56	
63		63
	71	
		80
	89	
		100
	112	
125		125
	141	
		160
	178	
		200
	224	
250		250
	282	
		315

TABLE A.2 (*Continued*)

1/1 *Octave band centre frequencies*	*Limiting frequencies*	1/3 *Octave band centre frequencies*
	355	
		400
	446	
500		500
	563	
		630
	709	
		800
	890	
1000		1000
	1122	
		1250
	1412	
		1600
	1780	
2000		2000
	2240	
		2500
	2820	
		3150
	3550	
4000		4000
	4460	
		5000
	5620	
		6300
	7090	
8000		8000
	8901	
		10000
	11220	

TABLE A.3

ABSORPTION COEFFICIENTS OF SOME COMMON MATERIALS[15]

	Frequency in Hz					
	125	250	500	1000	2000	4000
Brick, unpainted	·03	·03	·03	·04	·05	·07
Concrete, unpainted	·01	·01	·02	·02	·02	·03
Tiled floor, solid backing	·02	·03	·03	·03	·03	·02
Parquet floor	·04	·04	·07	·06	·06	·07
Wood joist floor	·15	·11	·10	·07	·06	·07
Plate glass window	·18	·06	·04	·03	·02	·02
Normal glass window	·35	·25	·18	·12	·07	·04
Plasterboard, 12.5 mm on timber frame	·29	·10	·05	·04	·07	·09
Plastered brick	·01	·01	·02	·03	·04	·05
Acoustic plaster, 25 mm	·25	·45	·78	·92	·89	·87
Fibreboard, 12·5 mm on solid backing	·05	·10	·15	·25	·30	·30
Mineral, glass wool, 25 mm solid backing	·15	·35	·70	·85	·90	·90
Same faced with 5% perforated hardboard	·10	·35	·85	·85	·35	·15
Carpet on good underlay	·08	·24	·57	·69	·71	·73
Curtains, heavy draped	·07	·31	·49	·75	·70	·60
Water surface	·01	·01	·01	·01	·02	·02
Audience, m² per person	·18	·40	·46	·46	·51	·46
Hard seats, unoccupied, m² per seat	·07	·10	·15	·17	·18	·20
Upholstered seats, unoccupied, m² per seat	·12	·20	·28	·30	·32	·37

References

1. Furrer, W., *Room and Building Acoustics and Noise Abatement*, Butterworths, 1964.
2. Meister, F. J., 'Protection against traffic noise', *VDI Zeitschrift* 1964, **106**, 23.
3. Maekawa, Z., 'Noise reductions by screens', *J. App. Acoust.* 1968, **1**, No. 3.
4. Kaye, G. W. C. and Laby, T. H., *Tables of Physical and Chemical Constants*, Longmans, Green, 13th Ed. 1966.
5. B.S. 3383:1961, I.S.O. R 266.
6. B.S. 3045:1958, I.S.O. R 131.
7. B.S. 4198:1967, I.S.O. R 532, Stevens.
8. B.S. 4198:1967, I.S.O. R 532, Zwicker.
9. Kryter, K. D., *J. Acoust. Soc. Am.* 1959, **31**, No. 11.
10. Kosten, C. W. and Van Os, G. J., 'Community reaction criteria for external noise', *Proc. Conf. Control of Noise*, H.M.S.O. London, 1962.
11. B.S. 4142:1967.
12. Noise. Final report by the Wilson Committee. H.M.S.O. Cmmd. 2056, 1963.
13. Langdon, F. J. and Scholes, W. E., 'The traffic noise index: a method of controlling noise nuisance', Building Research Station CP 38/68.
14. B.S. 3593:1963, I.S.O. R 266.
15. Doelle, L. L., 'Acoustics in architectural design', N.R.C. Canada, 1965.

Index

151